IMAGES
of America

INDIANAPOLIS TELEVISION

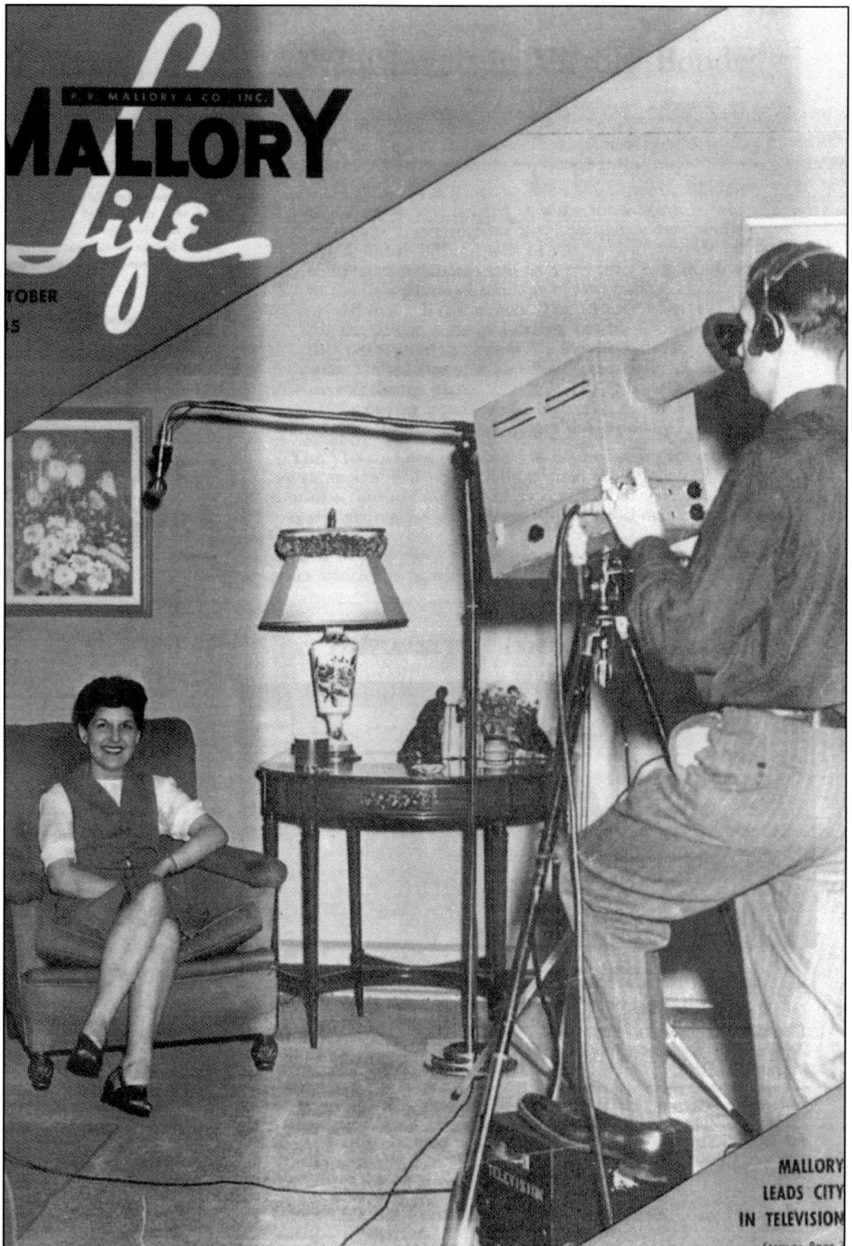

Here, Jerry Smith focuses the camera on Jane Gluff, a P.R. Mallory employee. The year was 1945, and the studio was in the Smith home on West Thirtieth Street. Mallory purchased several TV sets and placed them in the homes of Mallory employees. There was also contact with owners of about 31 other television receivers in the Indianapolis area.

ON THE COVER: Frances Farmer is on the set of her program *Frances Farmer Presents*. A stage and screen actress, Farmer was best known for the sensationalized accounts of her life. She was on stage, touring in 1958, when WFBM TV general manager Eldon Campbell saw her performance and offered her a job hosting afternoon movies. Farmer accepted, and her afternoon movie garnered top ratings throughout its run, ending in 1964.

IMAGES
of America

INDIANAPOLIS TELEVISION

David L. Smith

ARCADIA
PUBLISHING

ISBN 978-0-7385-9355-5

Published by Arcadia Publishing
Charleston, South Carolina

Printed in the United States of America

Library of Congress Control Number: 2012932290

For all general information, please contact Arcadia Publishing:
Telephone 843-853-2070
Fax 843-853-0044
E-mail sales@arcadiapublishing.com
For customer service and orders:
Toll-Free 1-888-313-2665

Visit us on the Internet at www.arcadiapublishing.com

This book is dedicated to the television pioneers.

Contents

Acknowledgments

The vast majority of photographs in this book are from the Ray Conolly collection. A special thanks goes to his son Jack for allowing us to use his father's photographs.

The following people also contributed photographs and/or additional material for this book: Dick Baldwin, Sue Bartlett, Natalie Bruner, Angela Cain and Young-Hee Yedinak (WTHR TV), Howard Caldwell, Janet Langhart Cohen, Jim Culbertson, Donald Davidson, Bev and David Fauvre, Mark Gramelspacher (CMW and P.R. Mallory Co., Inc.), Kevin and Bob Gregory, Charlie Haines, Scott Hainey (WISH TV), Ann Wagner Harper, Don Hein, Mary T. Holland, Bob Glaze, Clyde Lee and Diane Willis, Brad Lennon, Faith Levitt, Jimmy Mack, Chuck Marlowe, Bob McClain, Peggy McClelland, Carole McDaniel, Dawn Mitchell (the *Indianapolis Star*), Paul Montgomery (WRTV), Karen Mullen (WFYI TV), Biagio Presti, Lee Rosenthal (WXIN TV), Chuck Schisla, Carol Holland Sherburne, Terry Simpson, Bob Smith, Earl and Richard Sutherlin, Thomas Tarzian, Don Tillman, Earl Townsend III, Bob Warren, Dr. Darrell Wible, Jim Wilson, and Stan and Bobbi Wood.

INTRODUCTION

There were two migrations early in the nineteenth century. From the seaboard, there was a movement to the west. From the Carolinas and the mountain regions, there was a drift northward across the Ohio River. Indiana was settled by pioneers who had the enterprise to seek new fields and the gumption to unpack and settle down when they found themselves in the promise land.

This bit of philosophy by Hoosier writer George Ade summarizes a Midwestern attitude that is still prevalent. Ade could have added that the most sensible of the pioneers settled in the middle of the state and promptly dubbed their location "Indianapolis." The little city became a big city and is now the largest in the Midwest next to Chicago. Indiana and Indianapolis have achieved a peculiar distinction as a mixed type, a puzzling combination of people who are determined to do things their own way.

When television burst on the scene in the late 1940s, it brought forth trailblazers who had different visions. The William H. Block Company was one of the state's largest department stores, and it was the first to apply for a television channel. The company received it but had a change of heart and turned it back to the Federal Communications Commission (FCC). There were the Bitners, a father and son who believed there was a future for television. They stuck to their guns and made millions during the famous "TV freeze," which was when they were the only station in town. There was a Hoosier by way of Armenia, Sarkes Tarzian, who defiantly believed he could have a successful television station in the little town of Bloomington, Indiana.

Television had long been predicted. Most looked forward to the day when pictures and sound could travel through the ether. Finally, in 1927, a solitary inventor named Philo T. Farnsworth filed for his first patent of an electronic television system. Farnsworth later moved to Fort Wayne, Indiana, where he lived for 29 years after merging his company with the Capehart Corporation.

But there were only two applications granted for TV stations in Indiana before the FCC froze all other applications. One was in Indianapolis, and the other was in Bloomington. These two stations would be the only television stations in operation in the state for almost four years. While Sarkes Tarzian was going local with his programming in Bloomington, the Bitners in Indianapolis were picking and choosing programs from four national networks: NBC, CBS, ABC, and Dumont.

Tarzian had several very popular country music shows, but the most popular was *Hayloft Frolic*. Among the performers on that show was Bobby "Jingle Bell Rock" Helms. Local high school basketball and football games were televised, as was Indiana University basketball. There were variety and musical shows like *The Chandler Trio*, which featured Marvin Chandler and his sisters; *Leave It to George*, with humorist George Willeford; and Wally Bruner was "Wally the Weatherman" before he went on to become the host of *What's My Line?* The Indiana University Theater Department provided many dramas. Anything that was not on film was live.

The Bitners' station, WFBM TV, frequently inserted local shows into its network lineup. Viewers came to know and love local performers like Ann Wagner, weatherman Bill Crawford, newsman Gilbert Forbes, The Johnny Winn Trio, and Dick Lingle, as well as local high school basketball and coverage of the Indianapolis 500 Mile Race.

For much of its history, Indianapolis has been a manufacturing and industrial city. Today, however, the city has a much more diversified economy. Heavy emphasis is on education, health care, and finance. Tourism plays a growing role in the city. A newly expanded convention center is now one of the largest in the nation. The city is well known for hosting major sporting events. It is also the 25th largest television market in the nation.

One

The Experimental Years

May 16, 1939, is generally credited with being the date of the first telecast in Indiana; however, an engineering student at Purdue University may have beaten that date. The 1939 closed-circuit telecast was at the Antlers Hotel in downtown Indianapolis. Philco engineers set up a temporary installation as a demonstration of what the future might hold. It was reported that the images were clear but "tinged with a greenish-blue aura." In May 1940, as many as 23 stations were telecasting in the United States. But when America entered World War II, all plans to authorize more stations were held in abeyance.

In 1938, a ham radio operator named Jerry D. Smith received a TV receiver kit for Christmas. He built the receiver and then discovered he had nothing to look at. He decided to build a television camera. With the help of Purdue University, where he was an electrical engineering student, he managed to come up with a camera tube. Working with a fellow ham, Smith put together what was essentially a ham radio station (W9XBI) that was broadcasting television. Smith said it was not only the first TV station in Indiana but also the fifth station on the air in the world. The studio was in the Smith family home on West Thirtieth Street. Starting in 1939, the station was on the air every Monday with a test pattern, movie travelogues, and plays from Butler University. The station was operating experimentally on what would now be channel 2. Smith received an experimental license for the station in 1941. The P.R. Mallory Co. bought the station in 1944 and changed the call letters to W9XMT.

Jerry Smith was given the job of engineer in charge of the Mallory station. Marion Stevenson assisted Smith. To help in the experiment, Mallory arranged for the purchase of several television receivers that were placed in the homes of employees. According to Joseph E. Cain, vice president and general manager of Mallory, W9XMT would serve as a laboratory for development and testing of component parts for television. The station went off the air in 1947.

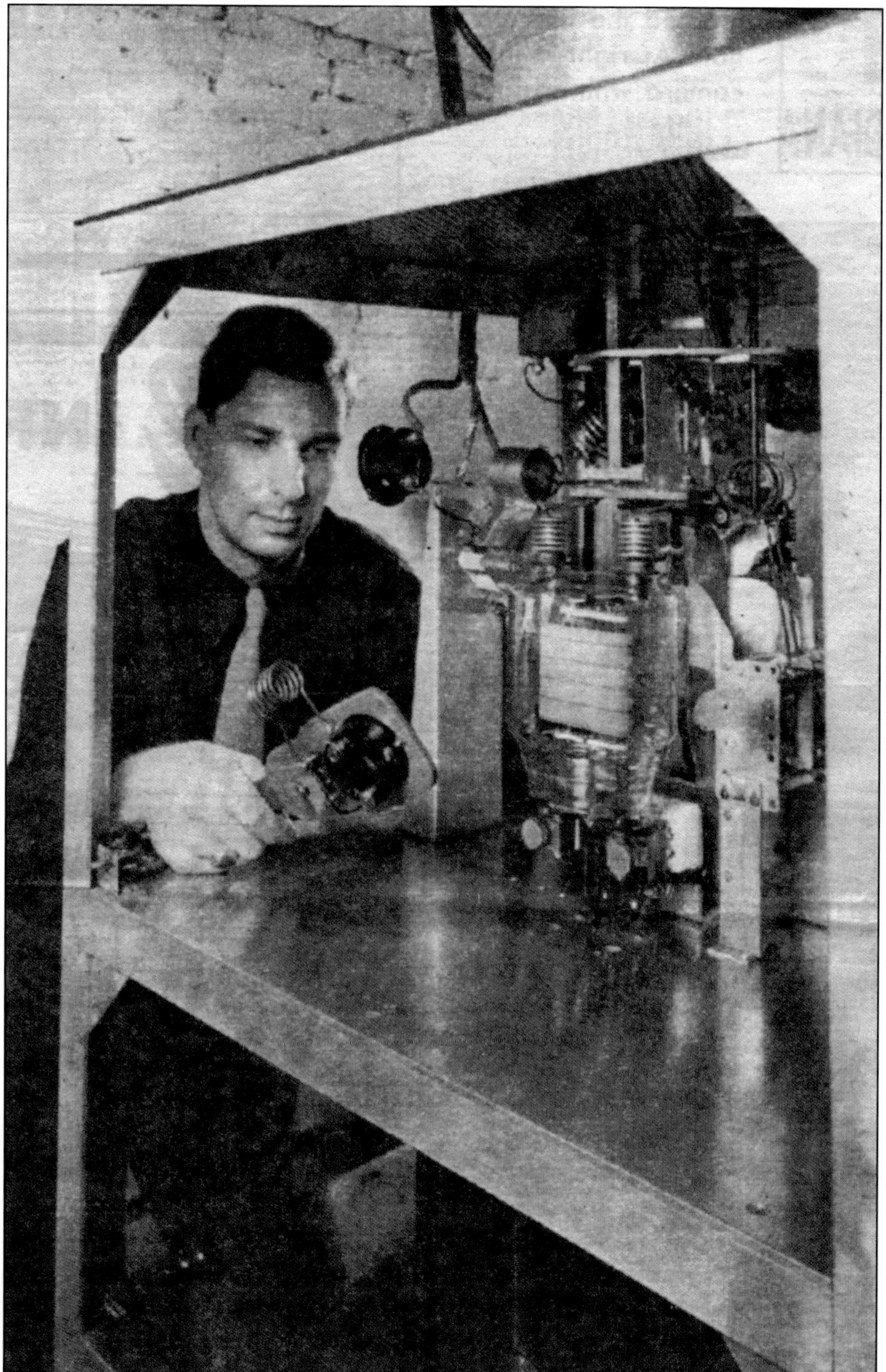

When Jerry Smith built this television transmitter in 1938, he had the only receiver in town. He was not sure how far the picture carried. By 1941, a few other receivers appeared in Indianapolis. Smith said, "There now were 11 sets. Sometimes we would have a camera pointed out on Thirtieth Street, and someone would call and say they had just seen a Ford drive by."

Harry Bitner Sr. was a newspaperman who ventured into radio and television. He purchased several stations throughout the country, including WFBM Radio in Indianapolis in 1939. WFBM TV signed on in 1949, and Bitner sold it, along with several other stations, to Time Life Broadcasting in 1957.

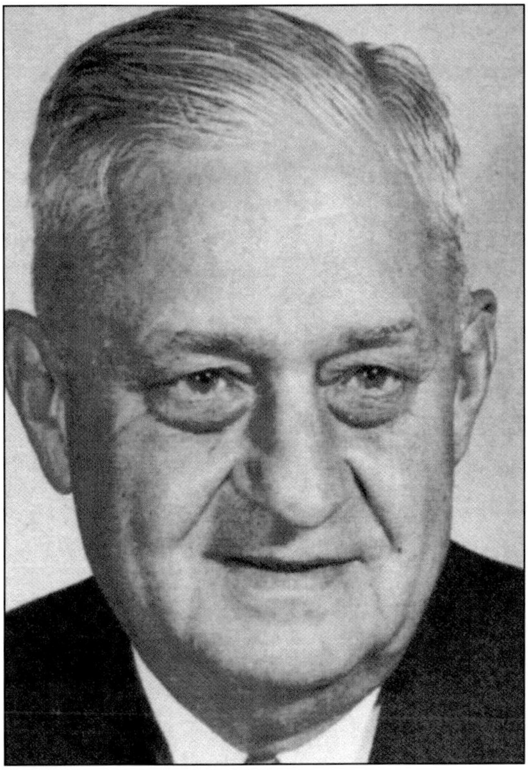

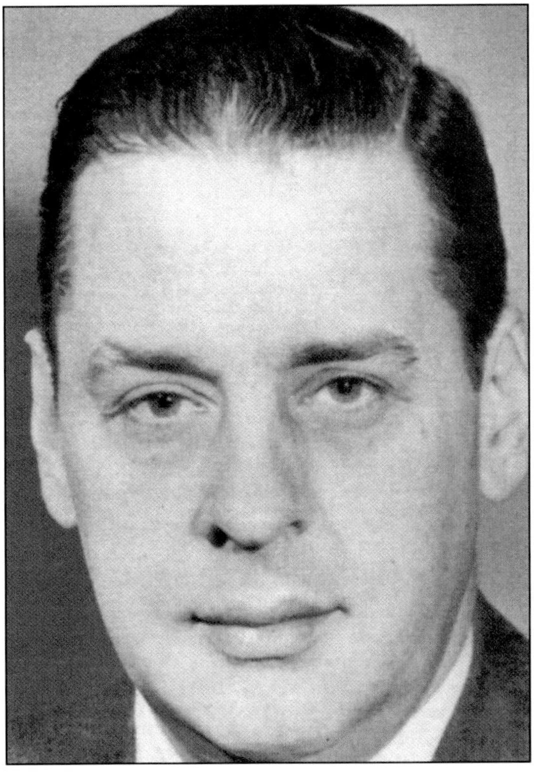

Harry Bitner Jr. became the general manager of WFBM Radio and oversaw the debut of WFBM TV. In September 1949, Bitner revealed the station was costing its owners $1,000 a day. It was then operating about 20 hours a week on a six-day schedule; it was off the air on Saturdays. In less than four months, the station had lost $57,000. The estimated number of television sets in Indianapolis at this time was 7,200.

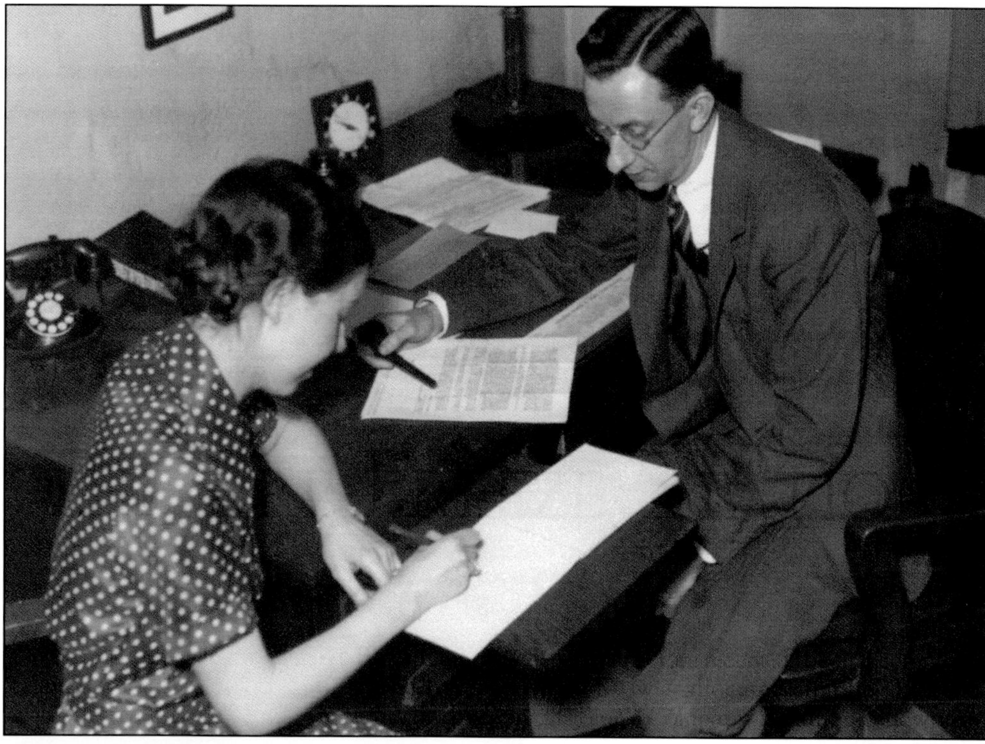

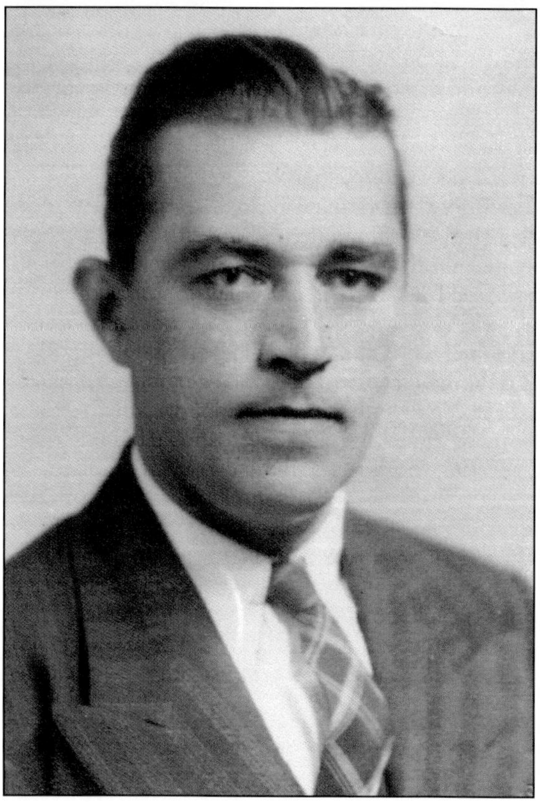

Frank O. Sharp was the longest tenured employee at WFBM. He was with the radio station when planning began in the early 1940s for the construction of a television station. At that time, Sharp was the station's acting manager. By 1943, the station engineers, under the guidance of chief engineer Harold Holland, had built a TV camera. When WFBM TV went on the air in 1949, Sharp was the program manager. He is shown with his longtime secretary Marthabel Geisler.

Harold S. Holland worked together with Frank Sharp as they turned their eyes toward the emerging world of television. Once it was decided that the first program would be the coverage of the Indianapolis 500 Mile Race, Holland and Sharp embarked upon a frenzied period of planning and building in order to meet the deadline of May 30, 1949. Holland and his engineers eventually built 10 cameras for studio and remote use.

Two

The Pre-Freeze Stations

On October 12, 1946, the Wm. H. Block Company department store received the first commercial television construction permit in the state. WWHB TV was allocated channel 3. Studios were located on one of the top floors of the Wm. H. Block Company Building in downtown Indianapolis. It bought equipment and began sending out test patterns. The planned airdate was to be in April 1948, but the company never got beyond sending out test patterns. It knew that most television stations at that time were losing money, and equipment was very expensive. The department store decided to forget about television and devote its full attention to its own business. The company voluntarily surrendered its license to the FCC, and the FCC moved channel 3 to Louisville, Kentucky.

By February 10, 1948, six applications for a TV station had been submitted to the FCC. WFBM TV, WTTV, and Block's Department Store had been granted construction permits. Other applications at this time were Indianapolis Broadcasting Co. (operator of WIRE Radio and owned by Central Newspapers), publisher of the *Indianapolis Star/News*, and the Crosley Broadcasting Corp. (operator of WLW in Cincinnati). These applications would not be acted upon for almost four more years.

By late 1948, the FCC had issued 108 television licenses, but problems were encountered, and the FCC declared a television freeze. The Korean War broke out in 1950, delaying the end of the freeze; finally, what began on September 30, 1948, ended on April 14, 1952. Only those who had been granted a construction permit before the freeze were allowed to go on the air. Indiana was fortunate to have two pre-freeze stations, WFBM TV (channel 6) in Indianapolis and WTTV (channel 10) in Bloomington.

The first commercial television station on the air in Indiana was WFBM TV. WFBM Radio had been purchased by Harry Bitner Sr. in 1939. Bitner was the publisher of the *Pittsburgh Sun-Telegraph* and former general manager of Hearst Newspapers. Harry "Pete" Bitner Jr. was named general manager of the radio station and became the general manager of Indiana's first commercial television station in 1949. Tests were sent out on May 16, 1949. Everything was ready for the inaugural program on May 30, 1949.

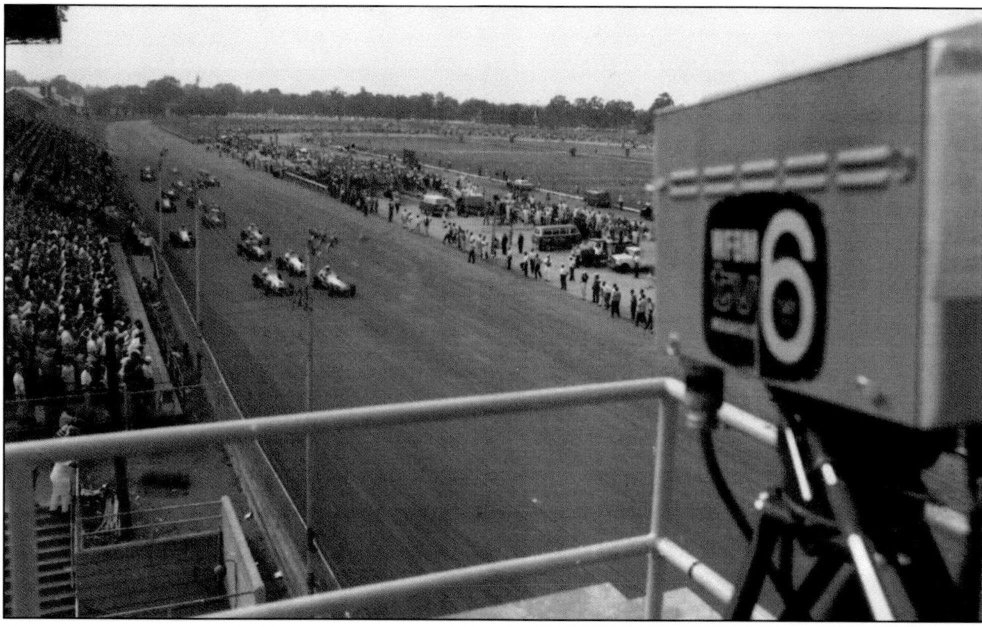

On race day, May 30, 1949, at 10:25 a.m., WFBM TV aired the first televised pictures of the racetrack of the Indianapolis 500 Mile Race. Popular WFBM disc jockey Paul Roberts, along with Mildred "Scottie" Fauvre (also with WFBM Radio), interviewed drivers and other celebrities in a makeshift studio under a grandstand.

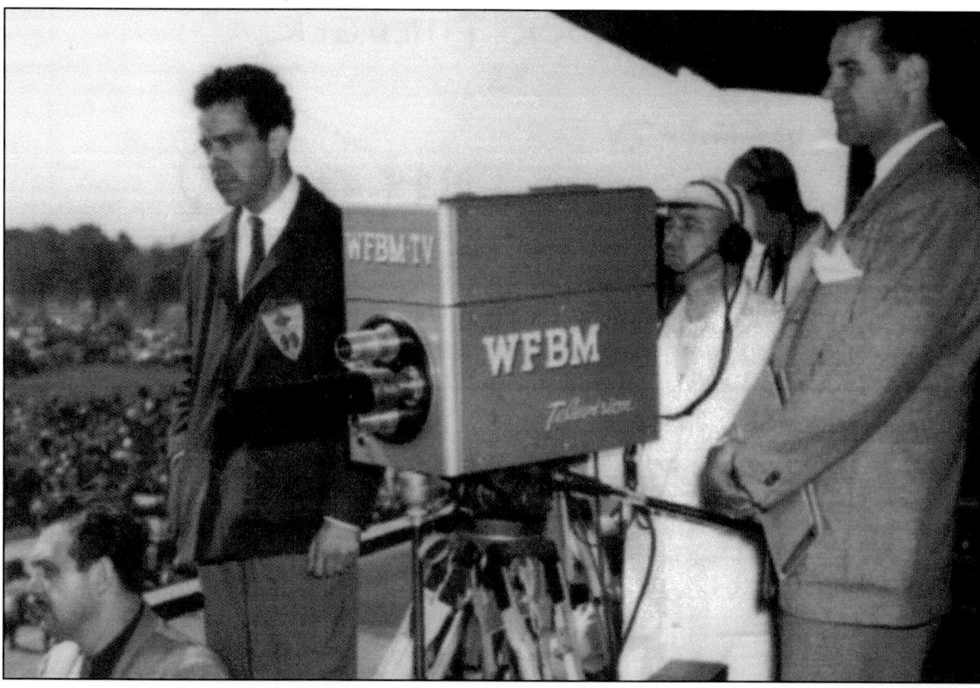

Earl Townsend Jr. did the lap-to-lap speedway coverage, while Dick Pittenger did the color commentary. Townsend eventually became a prominent attorney in Indianapolis, and Pittenger became WFBM Radio's primary sports personality. Pittenger is standing to the left of the camera, and Townsend is on the far right.

Paul Roberts became a nationally known disc jockey at WFBM Radio, where he was known as the "Happy Monster." From 1944 to 1950, he was the most colorful, unpredictable radio and television personality in Indianapolis. Unfortunately, he died of a heart ailment in 1950, just a year after making his television debut.

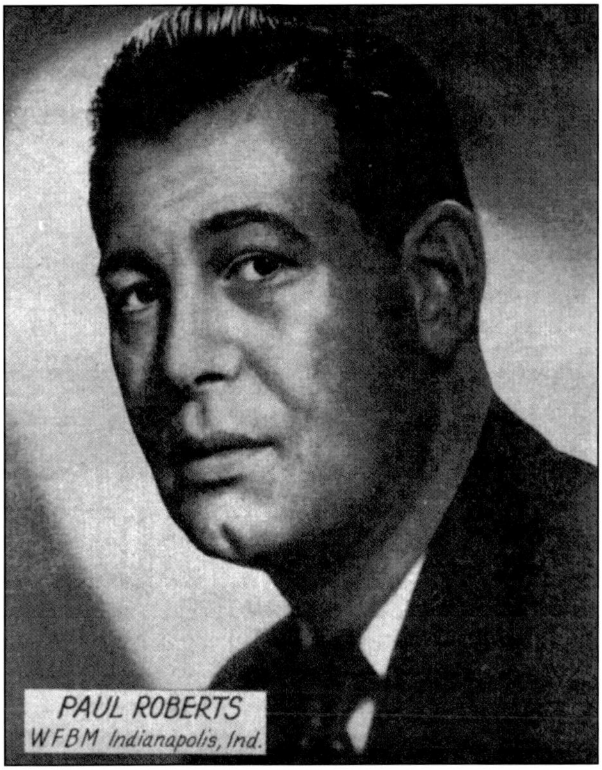

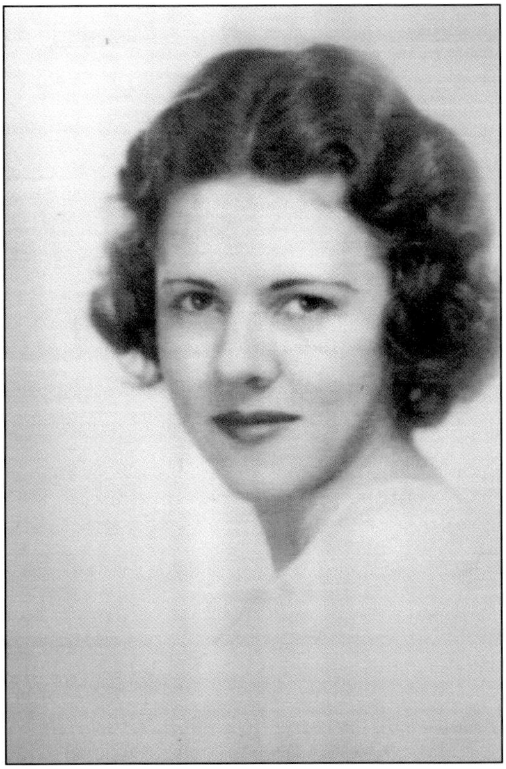

Mildred "Scottie" Fauvre joined WFBM Radio in 1941. She was a singer/actress and had her own 15-minute radio program entitled *Society Notebook*, where she reported society news and sang. Her husband, Irving, was a prominent attorney in Indianapolis. She had several firsts for a female broadcaster, including her participation in the first telecast of the Indianapolis 500 Mile Race.

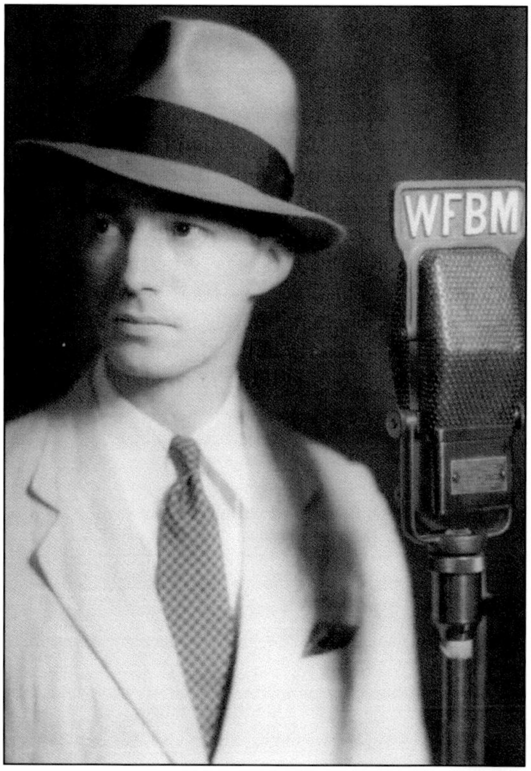

WFBM news director Gilbert Forbes wandered through the speedway crowd doing interviews. From 1937 to 1961, the voice of Forbes was the most respected and most welcome source of broadcast news in Indiana. He was a commentator, news editor, and analyst for WFBM Radio and, later, WFBM TV.

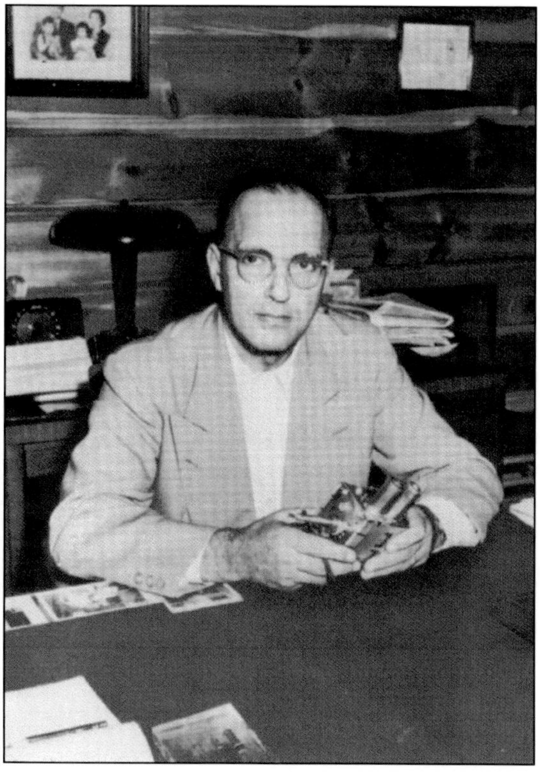

Sarkes Tarzian, a Hoosier by way of Turkish Armenia, envisioned a television station that would serve and be an integral part of the city of Bloomington, Indiana.

Bob Lemon was a guiding force behind the advent of WTTV in Bloomington. Tarzian had hired Lemon to help in sales. It was not long before Lemon became general manager of WTTV, and his success with the small-town station led him to Philadelphia in 1957, where he became program director of NBC-owned WRCV TV. In 1965, he moved to WMAQ TV in Chicago as vice president and general manager. In 1972, he was appointed president of the radio division of NBC.

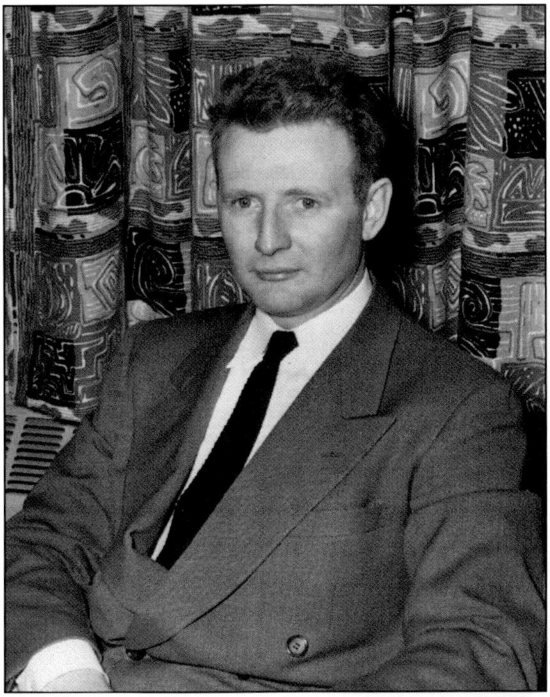

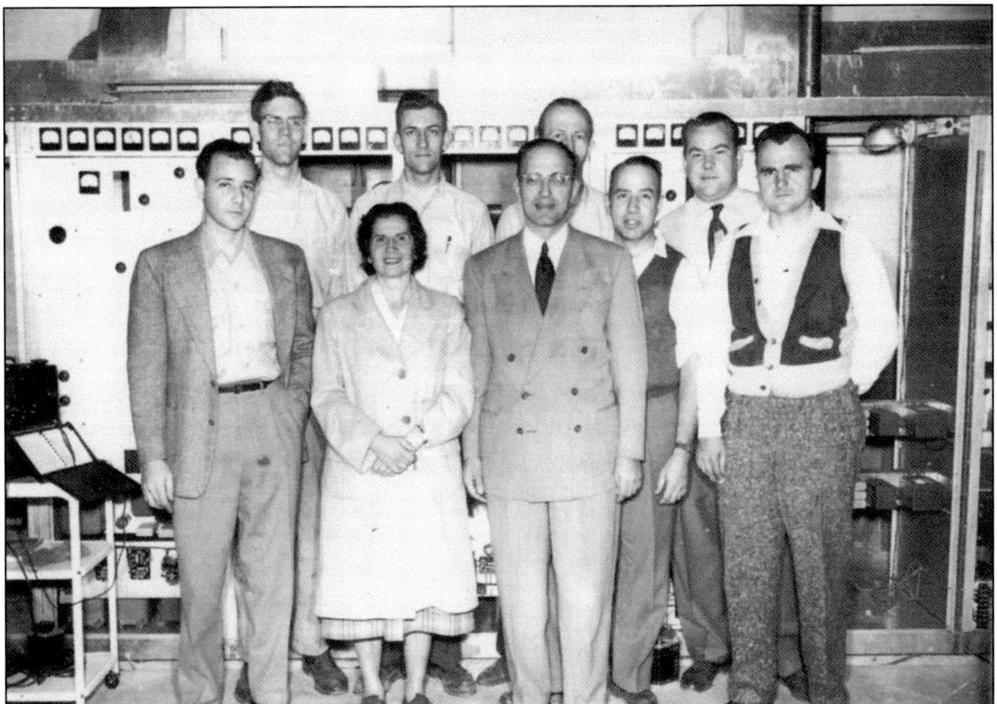

Standing in front of the channel 10 transmitter they built from scratch are Tarzian's engineers, who also built the camera and most of the equipment for WTTV's on-air debut. From left to right are (first row) Biagio Presti, Mary Tarzian, Sarkes Tarzian, George Morel, and Bob Young; (second row) Floyd Mallot, Carl Onken, Slim Fuller, and Mort Wiegel.

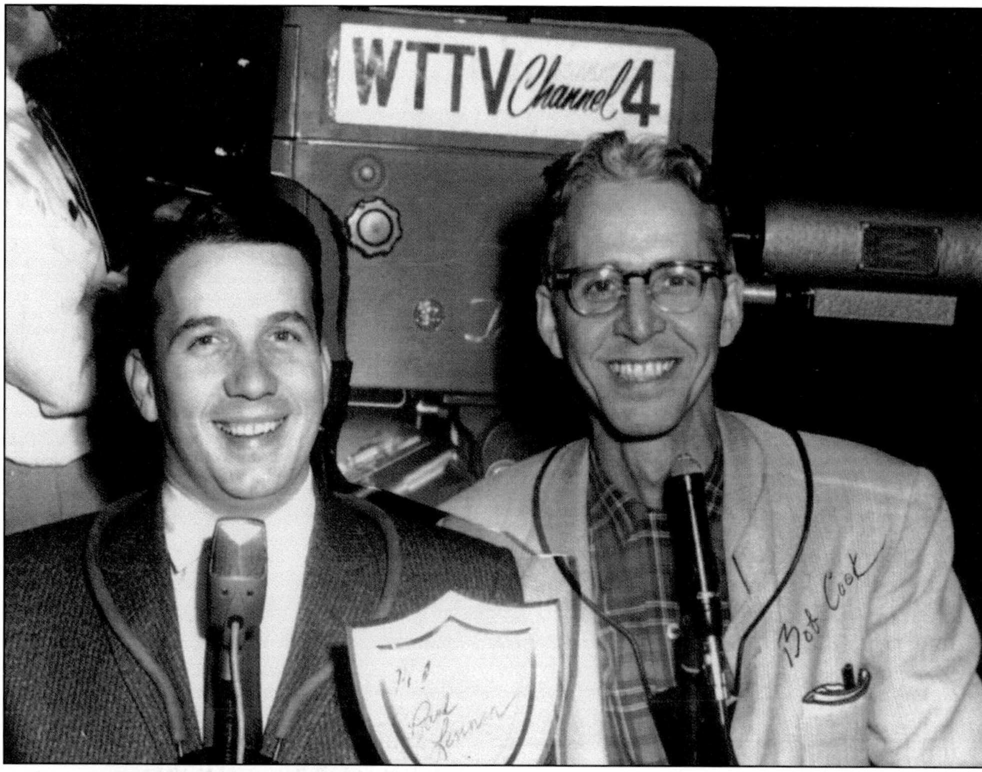

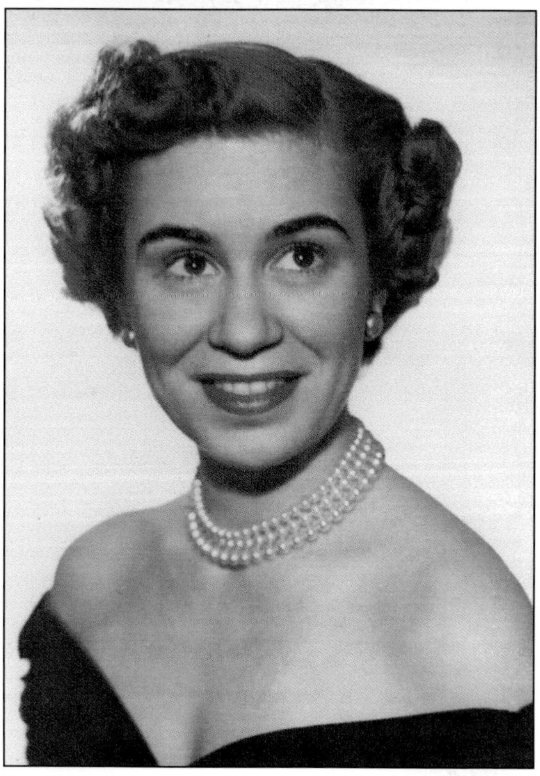

On December 6, 1951, Paul Lennon did play-by-play, and Bob Cook was the "color man" for the first regular-season college basketball game in the nation to be televised. Lennon was sports director and salesman at WTTV and was primarily responsible for convincing Chesty potato chips to be the sponsor of the Indiana University games. When Indiana University won the NCAA Championship in 1953, Chesty sales went through the roof.

Prior to the first telecast, WFBM had many promotional events to herald the coming of television. One of those events introduced a young lady who became an important part of local television history. On February 21, 1949, Ann Wagner was the featured attraction at a conclave of Admiral dealers at the Antlers Hotel. She appeared on closed-circuit television from the balcony of the hotel, and her image was on several 10- and 12-inch TV sets in the ballroom.

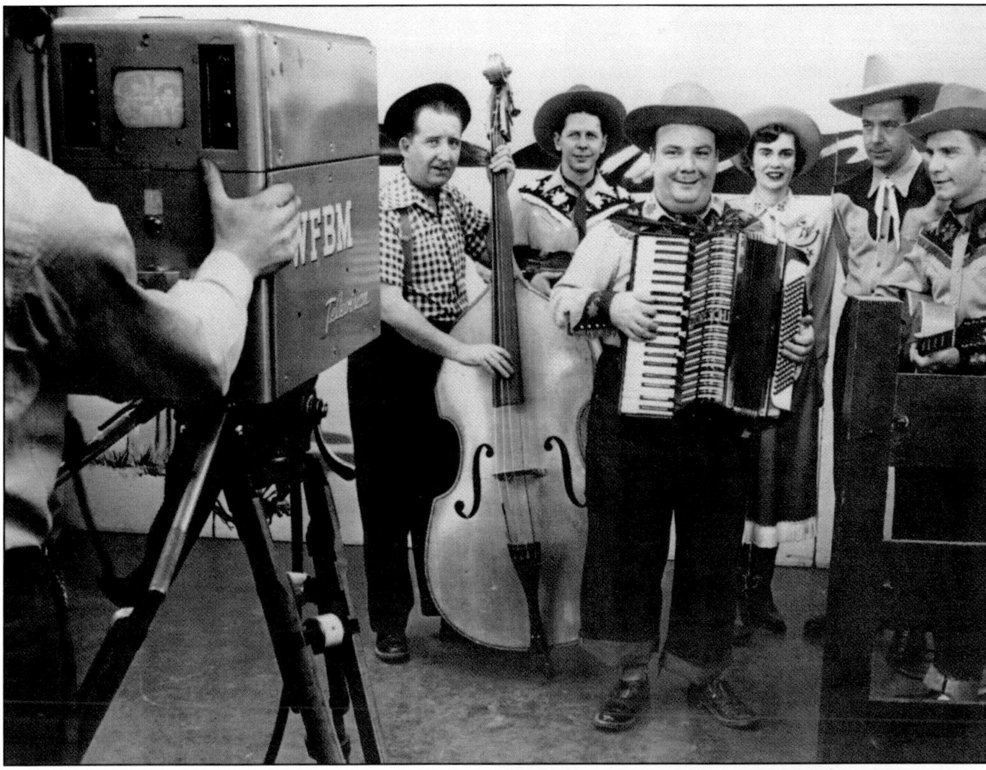

Ann Wagner was hired by WFBM Radio in 1947 before the television station went on the air. She was reputedly the first female disc jockey in Indiana and, likely, the first in the nation when she moved into popular disc jockey Paul Roberts's slot after his unexpected death. It was not long before she was asked to be a part of a new local television show, *Circle B Ranch*.

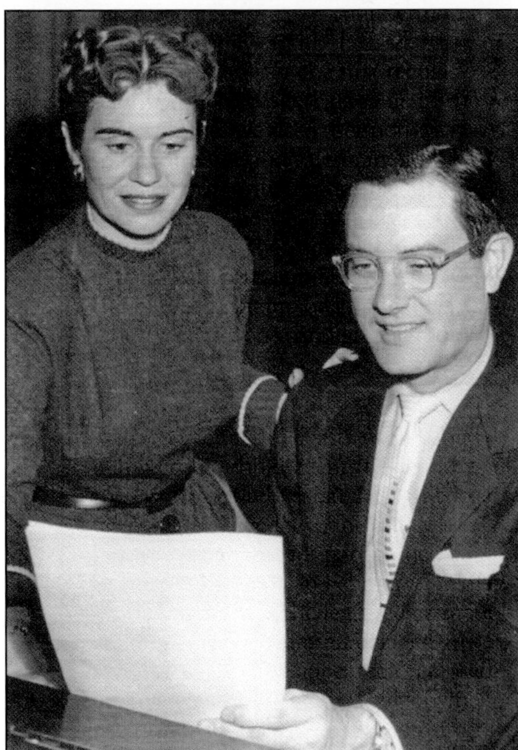

Ann Wagner soon became one of the busiest local personalities on early television. At one time, she was doing 13 TV shows a week. Here, she is with organist Kenny Jagger. She had another nightly show *Ann and Sue*, which featured Sue Dillman as her accompanist. In addition to her television shows, Wagner had two shows on WFBM Radio, *Ann's Pantry* and *Make Mine Music*.

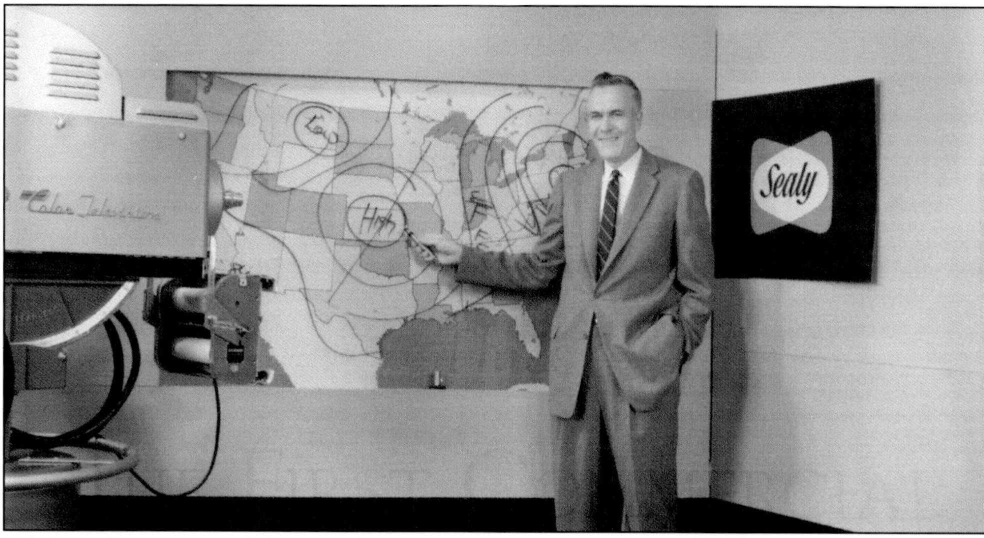

While auditioning candidates for the first weathercaster in June 1950, WFBM TV found a young dental student who seemed ideal. He had become a professional meteorologist in the military and had a nice mixture of humor, warmth, and integrity. Bill Crawford became a household name and continued as a weatherman after he received his dental degree and opened his practice.

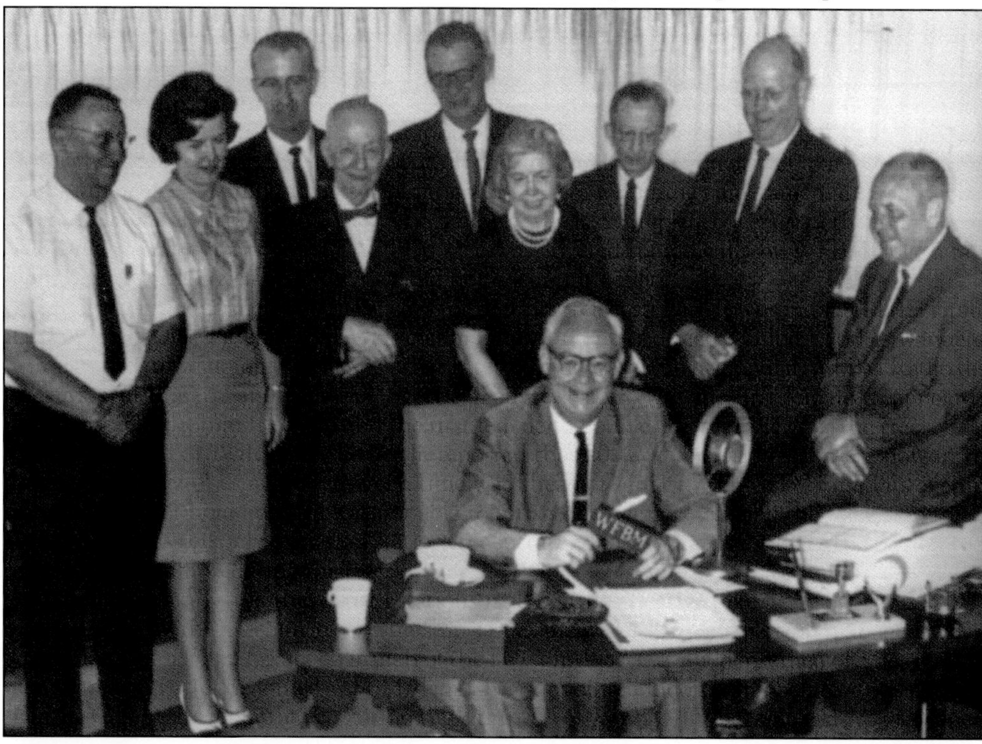

After the Bitners sold WFBM to Time, Inc., Wayne Coy, a Hoosier native and former chairman of the FCC, replaced Bitner Jr. as general manager of the stations. Coy brought another Hoosier with him. Eldon Campbell was in management positions at Westinghouse and Time, Inc., before returning to Indiana. In 1957, Coy died, and Campbell became vice president and general manager of the WFBM stations.

Three

The First Commercial Television Station

The inaugural program for WFBM TV began at 10:00 a.m. on May 30, 1949, with the film *Crucible of Speed*. At 10:25 a.m., the station switched to the first live pictures of the speedway. After the race coverage, the station went off the air, coming back on at 6:00 p.m. for four hours to air kinescope recordings of network programs. There were nine cameramen and technicians operating three cameras around the track. This was the last complete coverage of the Indianapolis 500 Mile Race for many years. The 1950 race was covered sporadically by WFBM due to rain. Full, live network coverage of the race did not occur until 1986.

Many people contributed to the success of that first telecast: Earl Townsend Jr. did the lap-to-lap coverage, Dick Pittenger did color commentary, and news director Gilbert Forbes wandered through the crowd doing interviews; other interviewing was done by popular WFBM Radio disc jockeys Paul Roberts and Mildred "Scottie" Fauvre. Fauvre was the first female to appear on broadcast television in the state and the first female to do live TV interviews at the track.

Near the end of 1951, Harry Bitner Jr. reported that 200,000 television sets were now in the range of channel 6. Demand for screen time was ahead of supply. WFBM TV raised its evening rates to $700 per hour and still was not able to satisfy the demand. During the television freeze, WFBM TV was the only television station in central Indiana. Sources estimate the station took in at least $6 million during the freeze.

The Bitner family sold the WFBM stations in 1956 (the effective takeover was May 23, 1957) to Time, Inc. With this change of ownership came a new president and general manager. Wayne Coy was a Hoosier and former chairman of the FCC. In 1957, Coy suddenly died of a heart attack less than a year after he joined WFBM. Time, Inc., appointed Eldon Campbell to Coy's position. During Campbell's tenure at the station (he left in 1974), WFBM TV became the broadcast news leader in the market. The combination of Bob Gamble as news director and Howard Caldwell as news anchor brought the station top news ratings for many years.

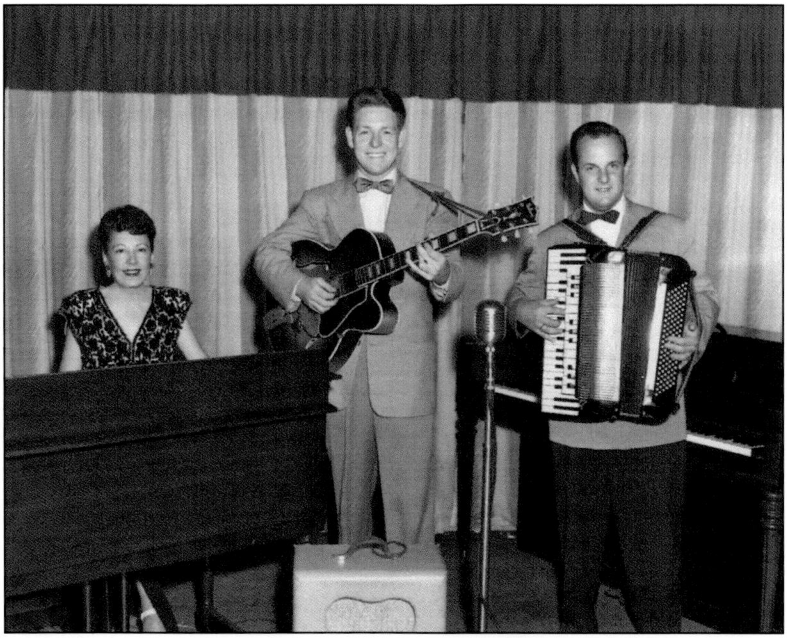

The Johnny Winn Trio was a very popular group in the early 1950s on WFBM TV. They were on at various times around 6:00 p.m. In many instances, the group was on for only five minutes; the longest time period in which they performed was 15 minutes.

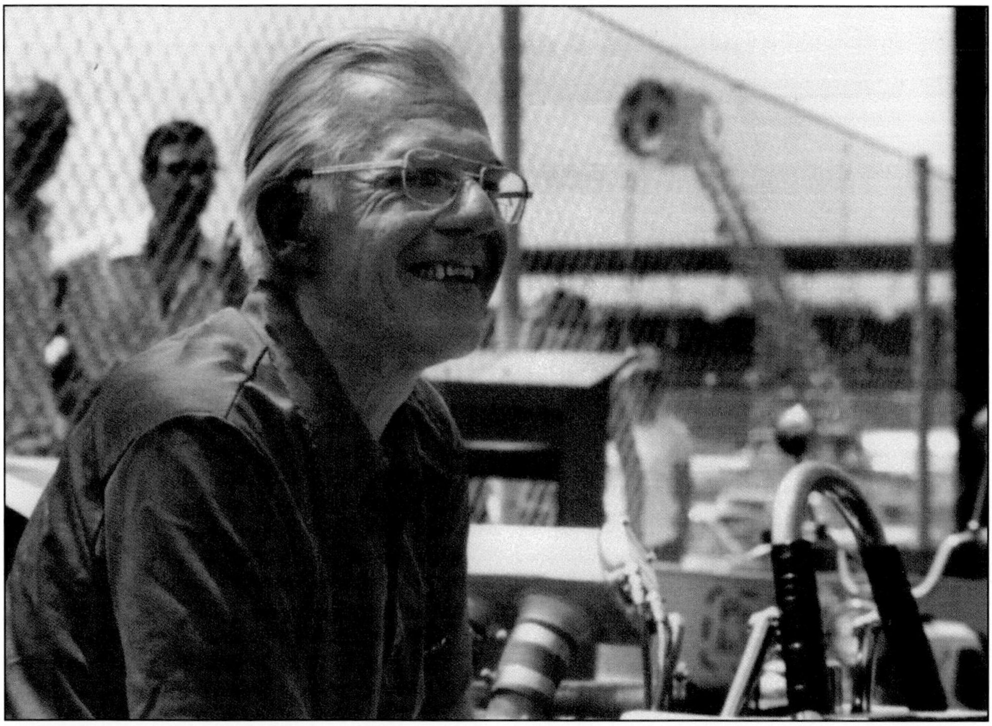

Tom Carnegie joined WFBM in 1953 and became a fixture on both radio and television. He joined the public-address team at the Indianapolis Motor Speedway in 1946. He retired from that job in 1985; by that time, he had become "the voice of automobile racing." Tony Stewart said, "You could take him to Daytona, you could take him to Sonoma, it doesn't matter. You put him on a microphone and if you know anything about racing, you know who that is at the other end of the microphone."

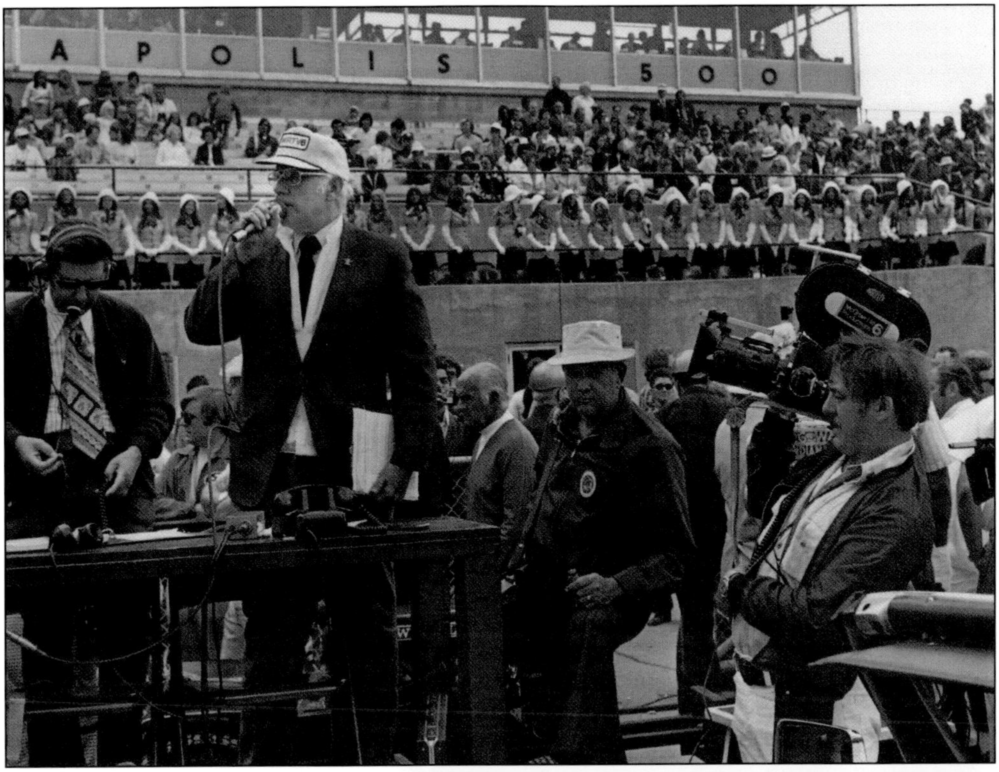

Carnegie was primarily identified with the speedway, but he covered all sports from the Indianapolis 500 Mile Race to high school and college basketball as well as football and golf tournaments. Here, he is seen at work with one of his daily *Trackside* programs during the month of May.

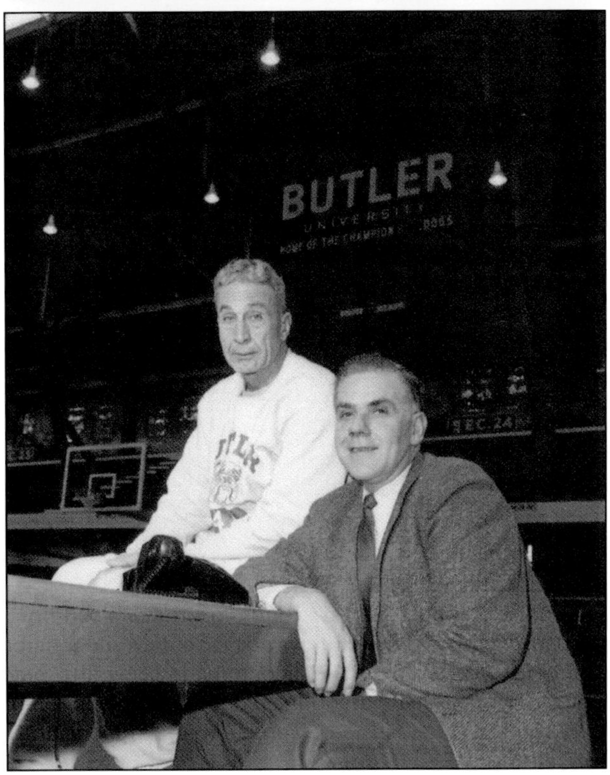

Tony Hinkle (left), the athletic director and coach of multiple sports at Butler University, teamed with Tom Carnegie to cover high school and college basketball.

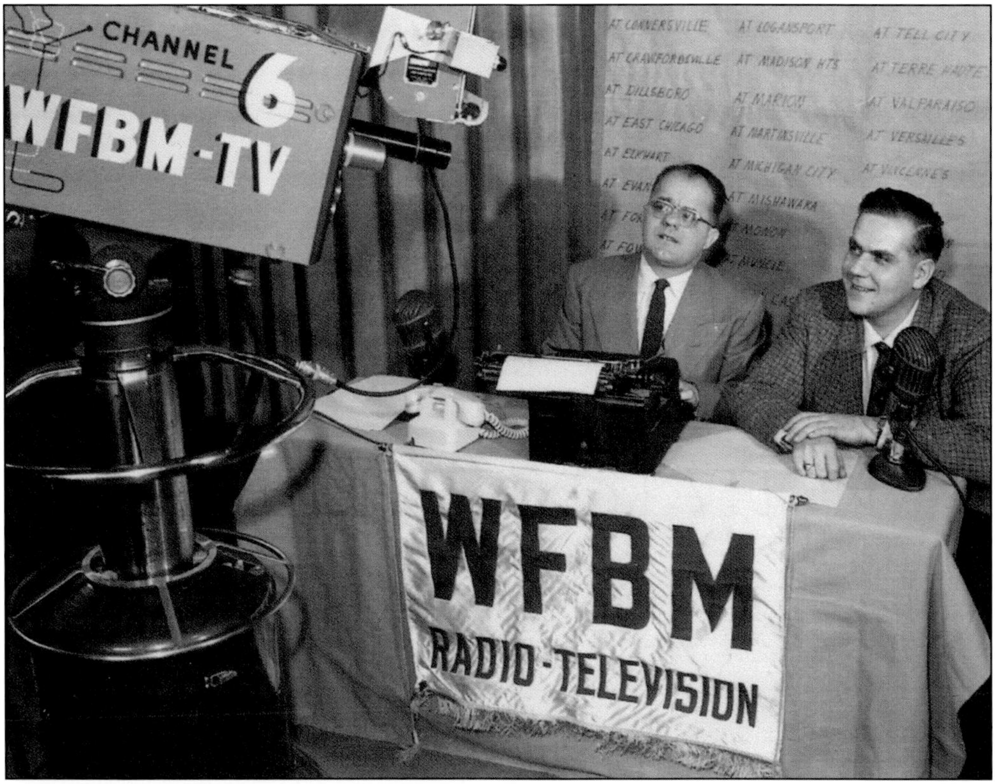

Tom Carnegie (right) is shown here with Jimmie Angelopoulos, sportswriter for the *Indianapolis Times*. These were the years that high school basketball was king, and any school could win the state championship.

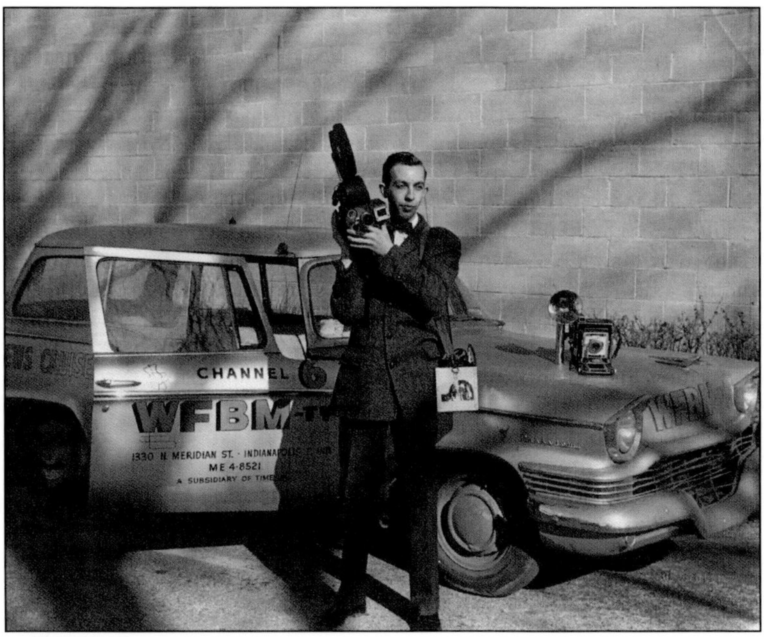

Dick Baldwin, one of the early photographers to be hired by WFBM TV, is ready to use his 16-milimeter-film camera. On the hood of the news unit is a Speed Graphic still camera.

Four

THE LITTLE STATION THAT COULD

Although Bloomington is about 50 miles south of Indianapolis, a Bloomington station became a factor in the Indianapolis market. The station was somewhat of an anomaly. Why did it try to exist in a town with a population of little over 28,000? The answer lies with Sarkes Tarzian, an emigrant from Turkish Armenia who was an electrical engineer. In 1940, he moved to Bloomington as chief engineer for RCA. In 1944, he founded Sarkes Tarzian, Inc., and opened a factory that manufactured TV tuners and selenium rectifiers. In 1946, he became owner of WTTS, a radio station in Bloomington.

The television industry was startled when Tarzian applied to the FCC for a channel for Bloomington. Most television experts wrote the venture off as an economic impossibility. Tarzian decided to entrust a group of his engineers with construction of the station. Mort Weigel, the leader of the team said, "Television was so new. Everything was different from anything we'd ever done before. Or anybody else for that matter. But everybody was so interested, we all just dug in."

After the freeze, the FCC changed a few channel numbers. WTTV was originally on channel 10, but the FCC assigned it channel 4, and channel 10 went to Terre Haute, Indiana. Since it was unlikely a coaxial cable would come to Bloomington, Tarzian built two relay towers between Bloomington and Cincinnati in order to get a network signal live from three Cincinnati stations. As other stations came on the air in Indianapolis, the networks affiliated on an exclusive basis with those who were legitimately Indianapolis stations. This meant that WTTV had no choice but to become an independent.

Tarzian eventually decided that his television station would not be able to survive in Bloomington. On February 21, 1954, a new tower was put in use in Cloverdale, Indiana, in the hope the station could cover both Terre Haute and Indianapolis; results were disappointing. In 1957, Tarzian successfully petitioned the FCC to move his tower to its current position at Trafalgar, Indiana. Tarzian sold WTTV in 1978, and it has had several owners since then. There was a time when Bloomington had the distinction of being the smallest city in the world with its own television station.

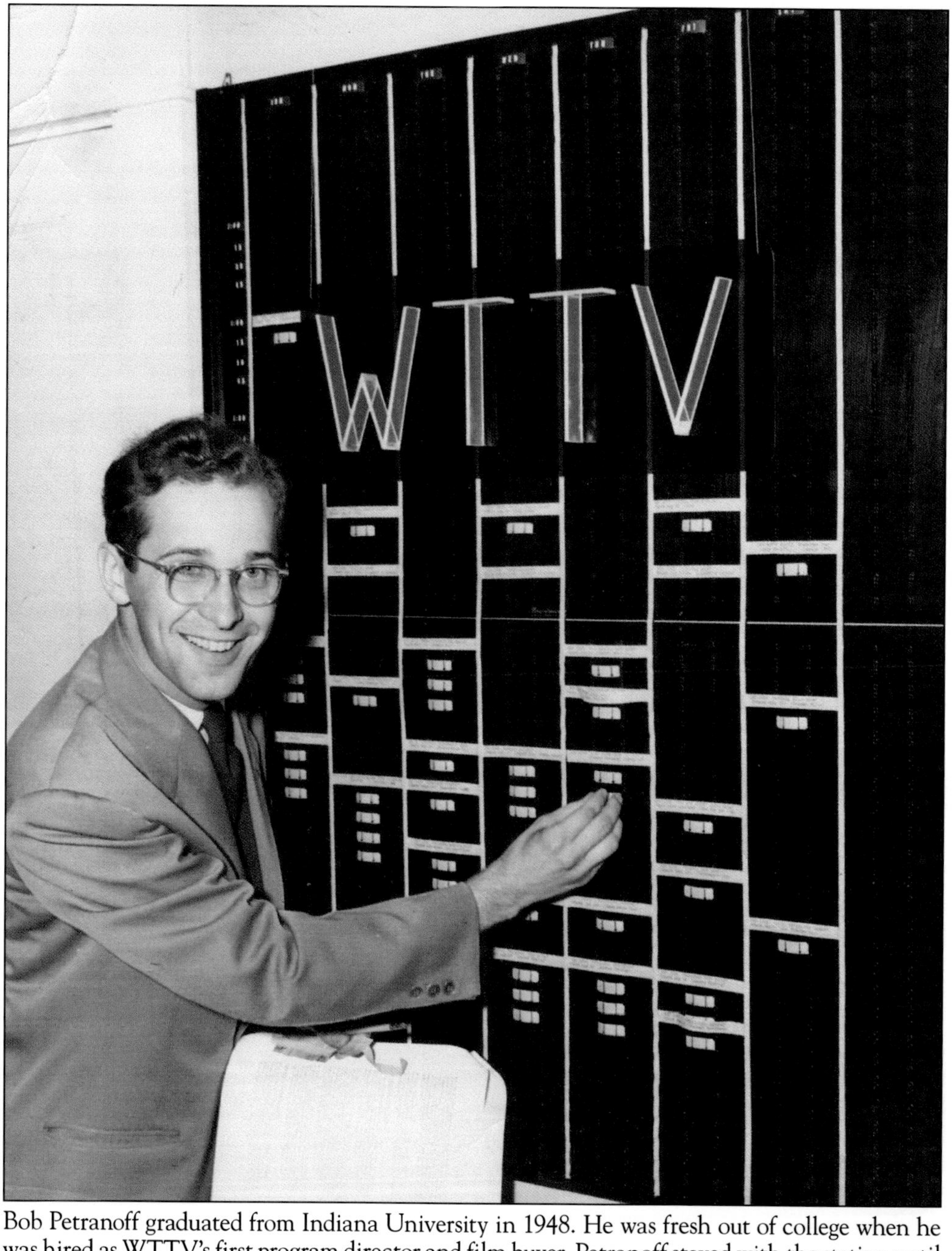

Bob Petranoff graduated from Indiana University in 1948. He was fresh out of college when he was hired as WTTV's first program director and film buyer. Petranoff stayed with the station until 1958, when he became director of Indiana University Radio and TV Services in Bloomington.

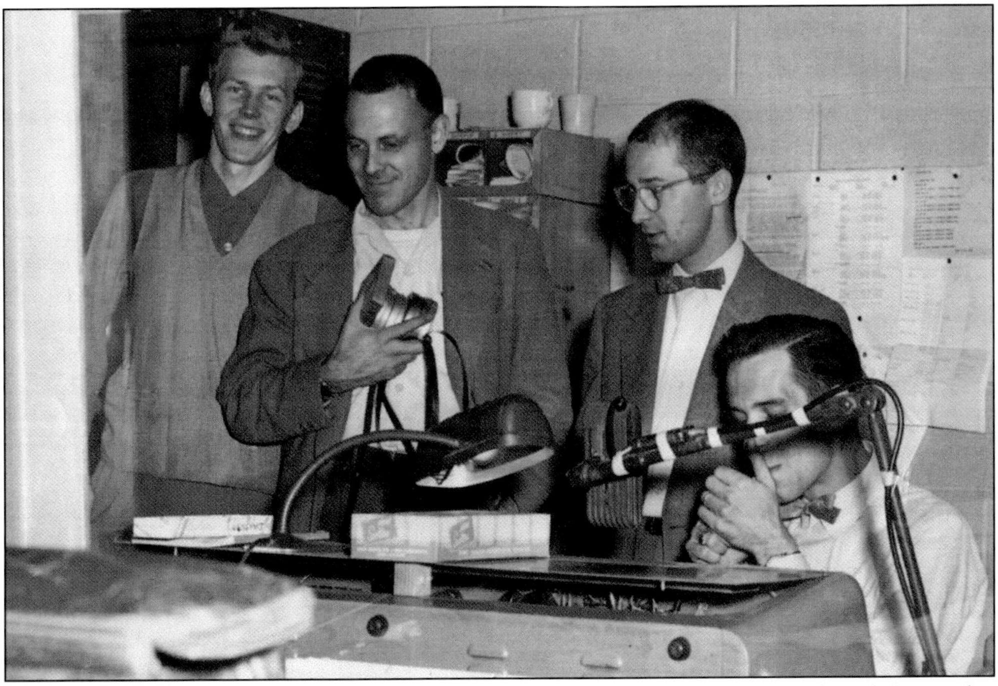

In the control room at the Hillside Drive studios in Bloomington in 1954 are, from left to right, Dave Smith, director; Warren Perney, film editor; Bob Petranoff, program manager; and Doug Brown, announcer.

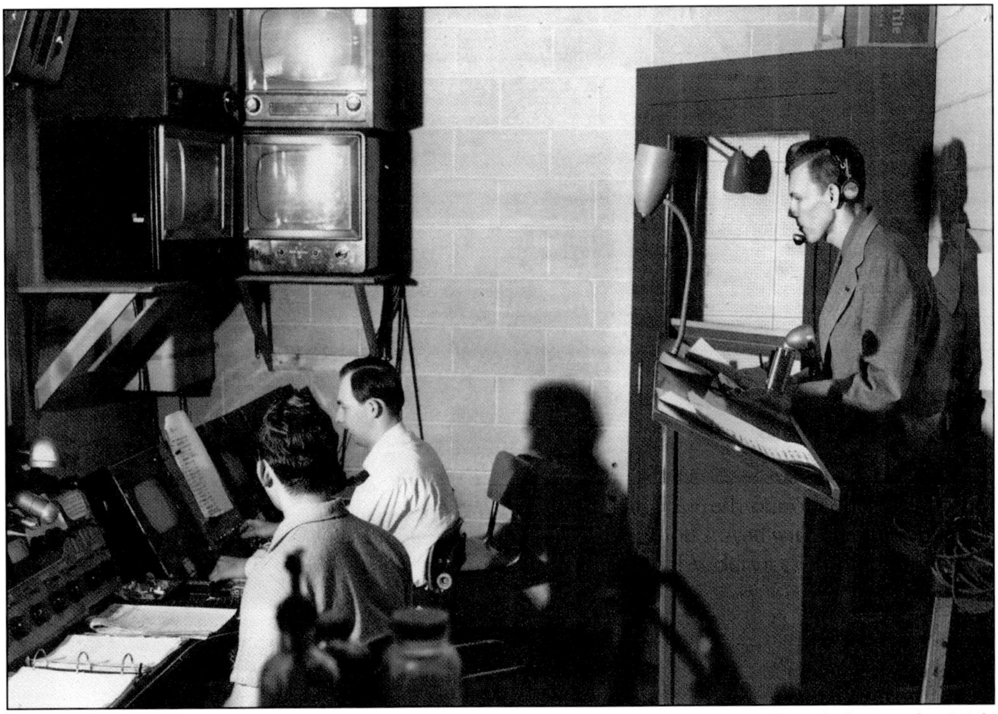

Monitoring a live production in progress at WTTV in the early 1950s are, from left to right, Milt Lewis, Bill Wheat, and Dave Smith.

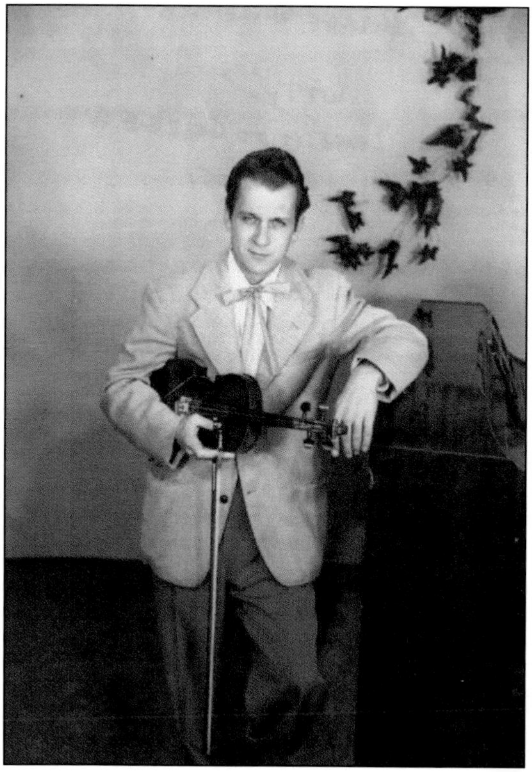

George Willeford was an instructor at Indiana University in the daytime and talent on WTTV with his own television program, *Leave It to George*, in the evening. Willeford later moved to channel 13 in Indianapolis, where he hosted movies and had his own interview show.

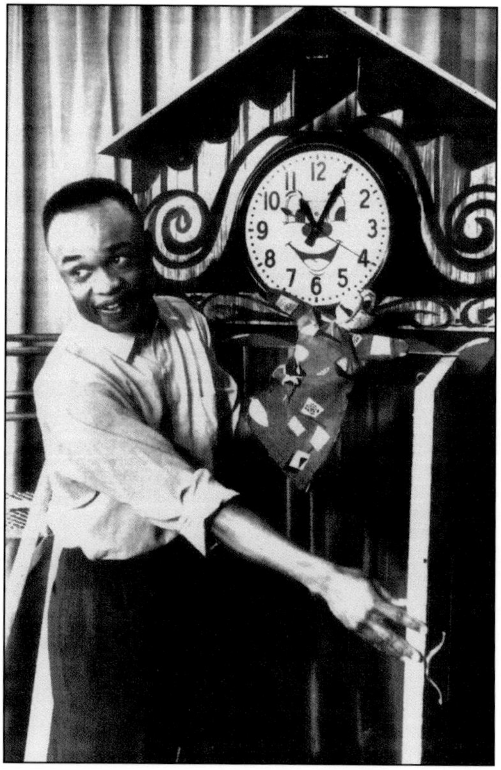

One of the first staffers to be hired at WTTV in 1949 was Charlie Haines. Haines was an Indiana University student and had been working at WTTS Radio part-time as an advertising artist. He moved to WTTV and became the one and only full-time artist at the station. When WLWI TV went on the air in 1957, Haines accepted an offer to become its chief artist. He was with WLWI/WTHR TV until 2006.

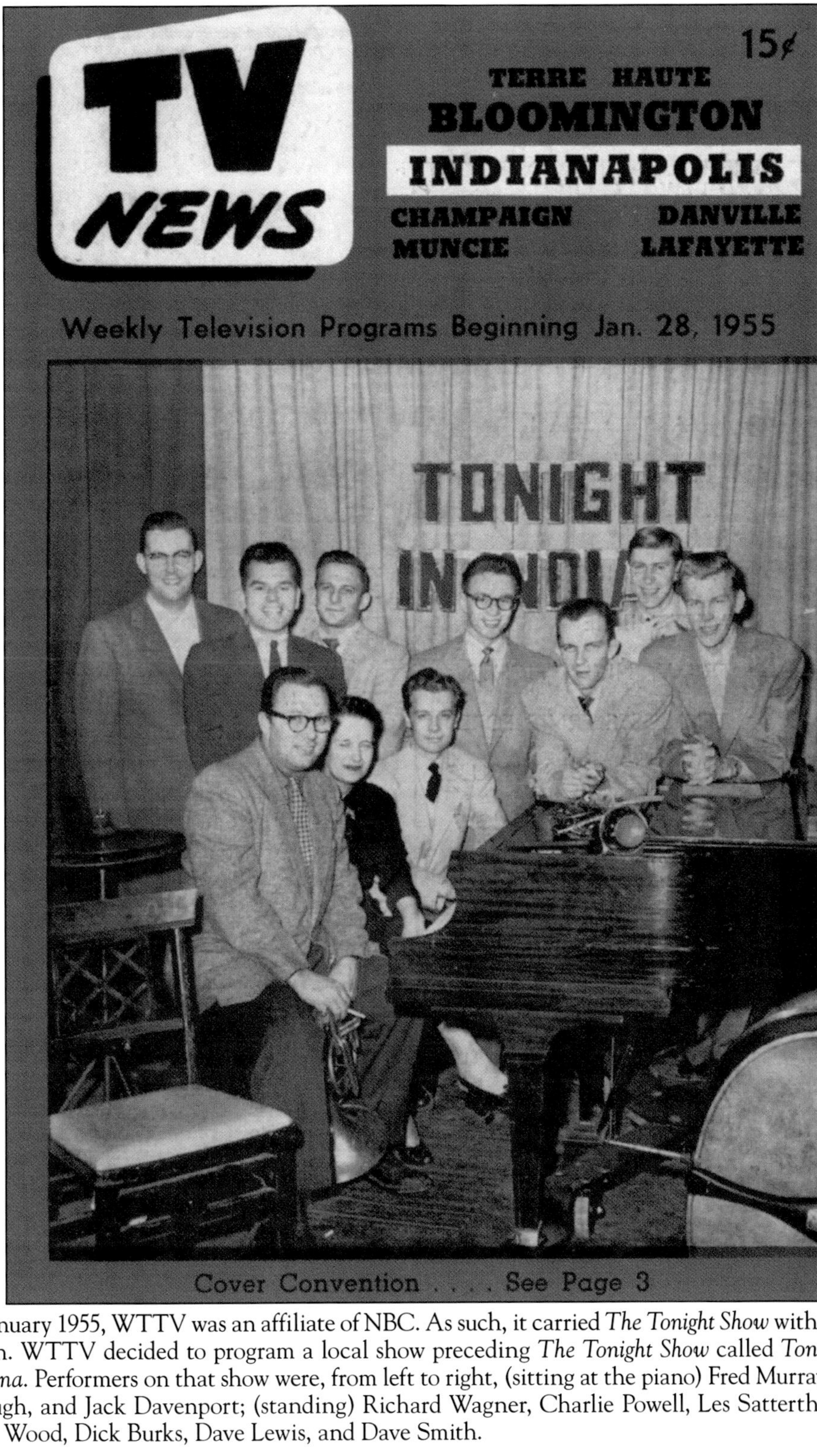

In January 1955, WTTV was an affiliate of NBC. As such, it carried *The Tonight Show* with Steve Allen. WTTV decided to program a local show preceding *The Tonight Show* called *Tonight in Indiana.* Performers on that show were, from left to right, (sitting at the piano) Fred Murray, Mid Keough, and Jack Davenport; (standing) Richard Wagner, Charlie Powell, Les Satterthwaite, Stan Wood, Dick Burks, Dave Lewis, and Dave Smith.

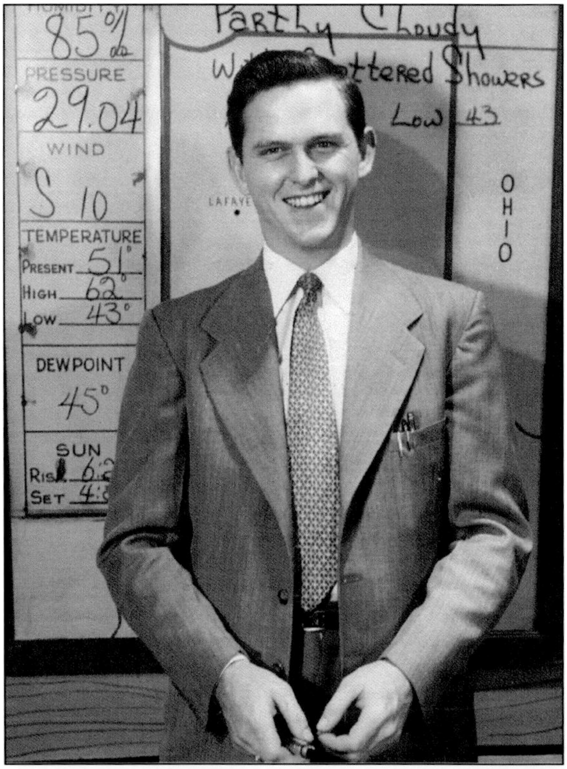

Wally Bruner was an early weatherman at WTTV. He attended Indiana University while working at the station. Many of the early staff members were Indiana University students who worked part-time. Bruner later became a news correspondent for ABC, reporting from Vietnam, Santo Domingo, and the White House during the Johnson administration.

Bruner left the news profession to become host of the syndicated *What's My Line?* television series. He later produced and hosted his own syndicated series, *Wally's Workshop*. He came back to Indiana in 1967, where he and wife, Natalie, hosted a daily talk show, *The Morning Show*, on WISH TV.

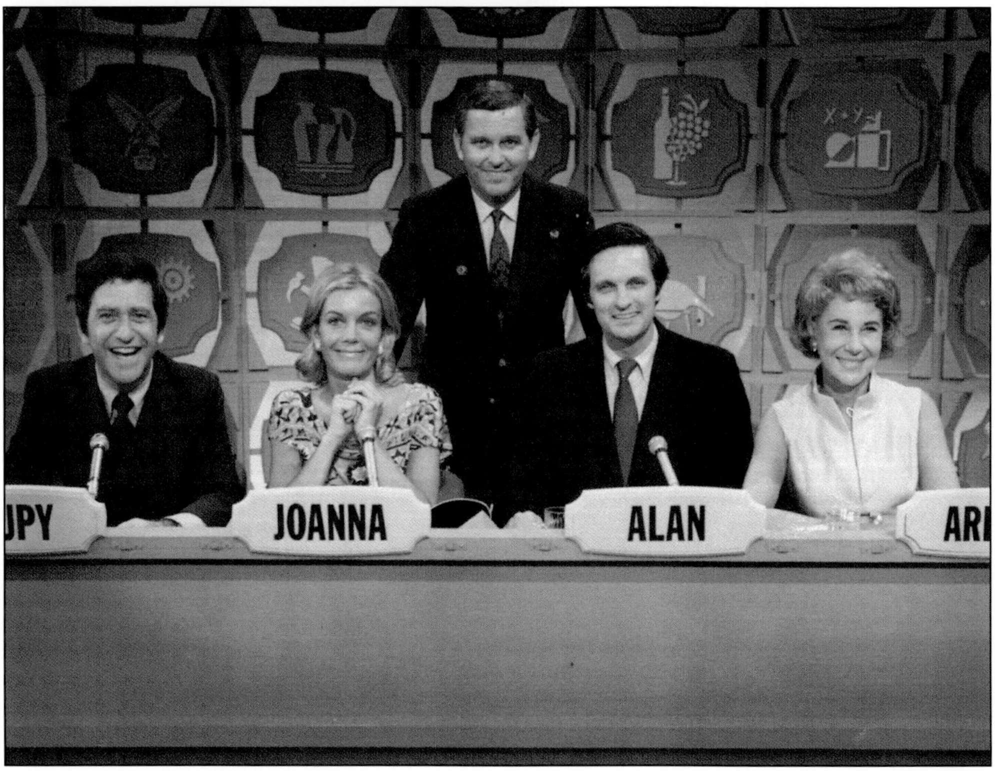

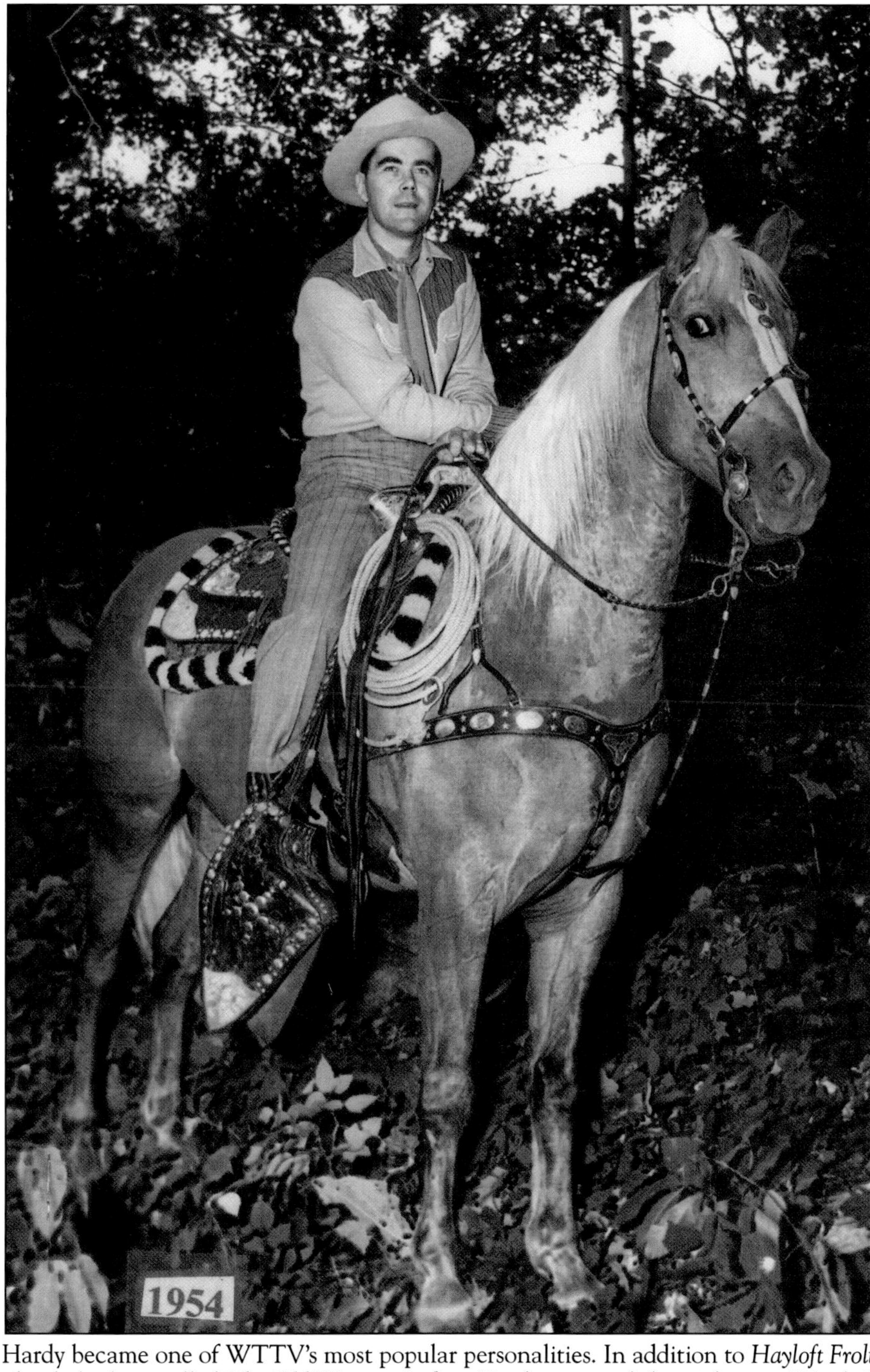

Bob Hardy became one of WTTV's most popular personalities. In addition to *Hayloft Frolic*, he hosted a daily show called *The Old Western Ledger*. Hardy acquired the nickname "Uncle Bob," as he introduced a western movie each day. He decided to form a Western Ledger Club, and it grew rapidly to 63,000 members. Hardy bought a Palomino horse and used it at the opening of the show and for personal appearances. When he staged a Western Ledger Member Roundup at McCormick's Creek State Park, 35,000 people showed up.

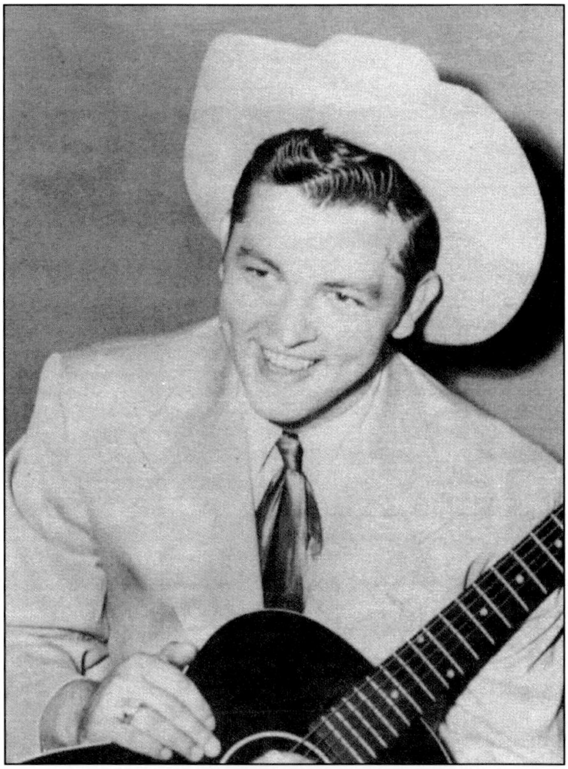

One of the most popular local shows on WTTV was Bob Hardy's *Hayloft Frolic*. One of the performers was a local Bloomington boy named Bobby Helms (pictured). Helms left for Nashville, where he recorded such hits as "Jingle Bell Rock" and "My Special Angel." Another performer who left for Nashville was Joe Edwards, becoming a guitar impresario. Bob Hardy was the emcee, producer, and star of the *Frolic*. His popularity with viewers was legendary.

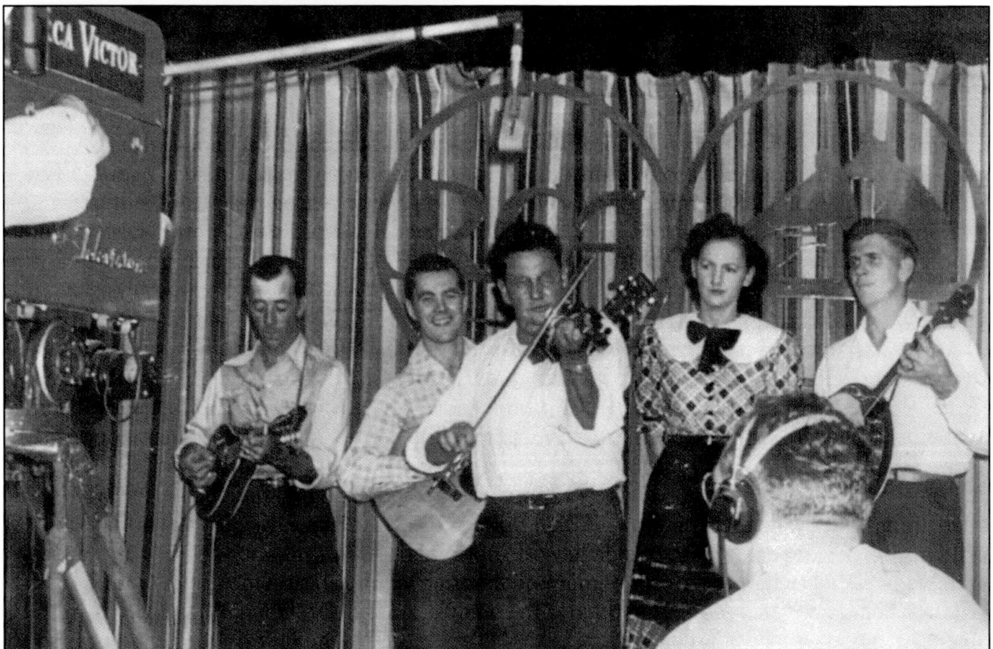

In 1947, RCA put on a demonstration of television. Bob Hardy brought his country music act and performed. It was an indication of things to come, and in two more years WTTV signed on, and Hardy became a fixture on the station. Hardy is standing behind the violinist.

Five

The Second Indianapolis Station

On March 15, 1952, an application for a construction permit for a second Indianapolis station was filed with the FCC. Television Indianapolis, Inc., had stockholders like L.S. Ayres & Co., Butler University, P.R. Mallory & Co., Inc., and radio station WXLW. Apparently, this application was withdrawn.

However, three months later on June 18, 1952, another application was filed with the FCC for a second Indianapolis-based television station. C. Bruce McConnell, president of Universal Broadcasting Company, Inc., and owner of WISH Radio, announced the station had been assigned VHF channel 8. Other officers and directors of Universal were Frank McHale, William Spencer, Frank E. McKinney, Earl Schmidt, Robert McConnell, and Stokes Gresham Jr. McConnell was also president of Hamilton-Harris & Co., one of the country's largest jobbers of tobacco and sundry products.

The tower was built at the WISH Radio transmitter site at Post Road and Rawls Avenue. A new three-story addition to the WISH Radio studio in the Riddick Piano Building at 1440 North Meridian Street was constructed and completed in time for the opening telecast. The station was officially opened with a dedicatory telecast on July 1, 1954. C. Bruce McConnell presided at the dedication, with Robert B. McConnell and Stokes Gresham Jr. participating. Robert McConnell, the son of C. Bruce, was the vice president and general manager of the station. Stokes Gresham Jr. was vice president and chief engineer.

The station started with a basic ABC affiliation (the network affiliate of WISH Radio), but as was the case in television's early days, the station also was able to carry some programs from NBC, CBS, and DuMont. A multitude of new national programs appeared as well as local personalities Virginia Byrd, Reid "Chuckles" Chapman, Gene Allison, Luke Walton, Dee Sweet, and newsmen Vince Leonard and John Fraim.

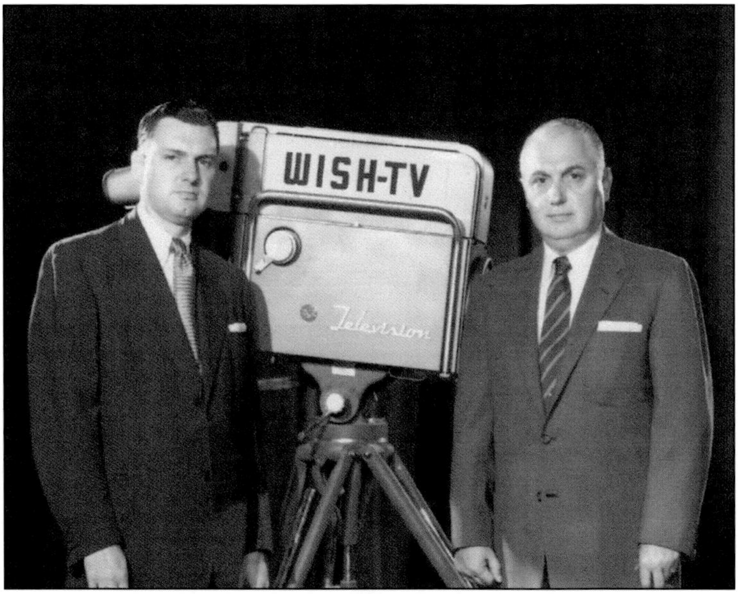

The father-son combination of Robert B. McConnell (left) and C. Bruce McConnell was the driving force behind the creation of WISH TV channel 8. The McConnells owned radio stations in South Bend, Anderson, and Fort Wayne. They were also the principal owners of WINT TV in Waterloo, Indiana, which later became WANE TV channel 15.

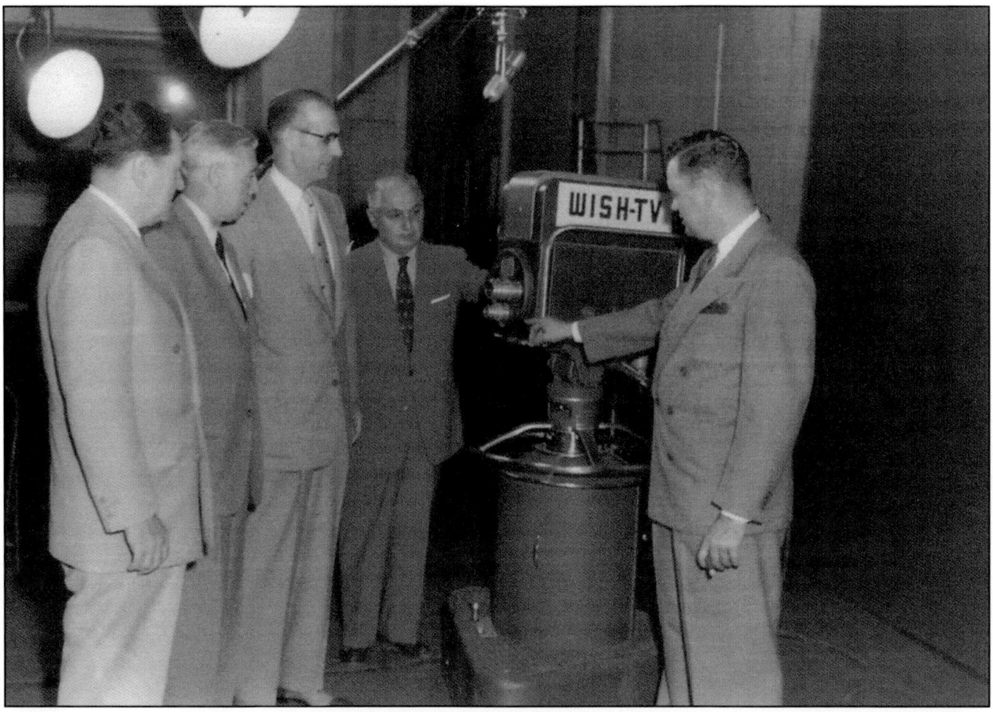

Prior to signing on the air for the first time, vice president and chief engineer Stokes Gresham Jr. explains the workings of a TV camera for stockholders, from left to right, Frank E. McKinney, William Spencer, Earl Schmidt, and C. Bruce McConnell. Gresham designed the studios for WISH Radio and TV, including a 150-seat auditorium studio. The auditorium was a designer's dream, featuring a saw-tooth ceiling and polycylindrical walls for the finest acoustics.

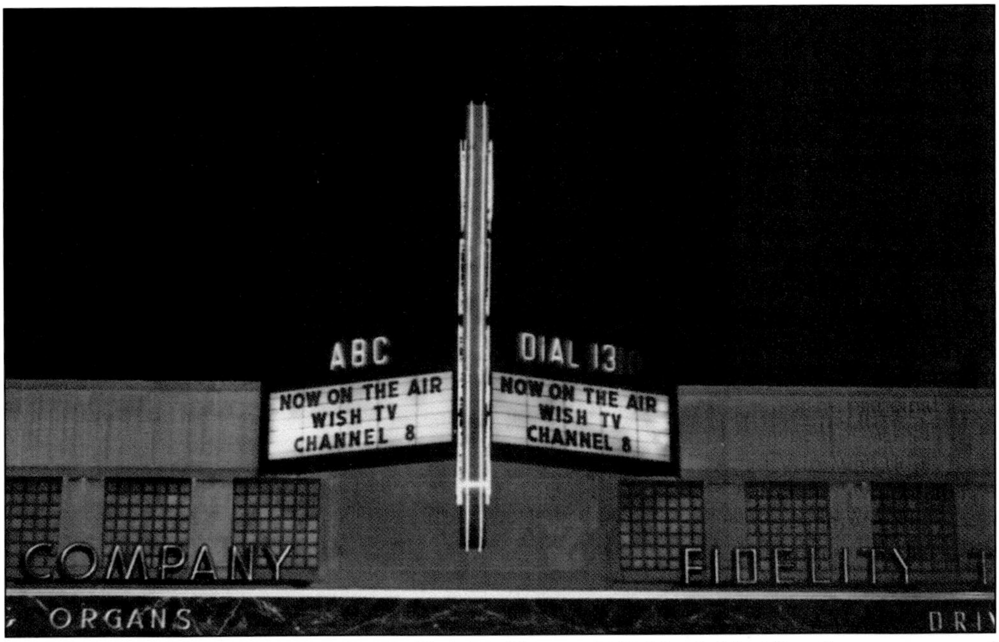

On July 1, 1954, the marquee on the Riddick Building at 1440 North Meridian Street trumpeted the fact that a second Indianapolis-based television station was now on the air. WISH TV and WISH Radio occupied the second and third floors of the building, which had been revamped for television.

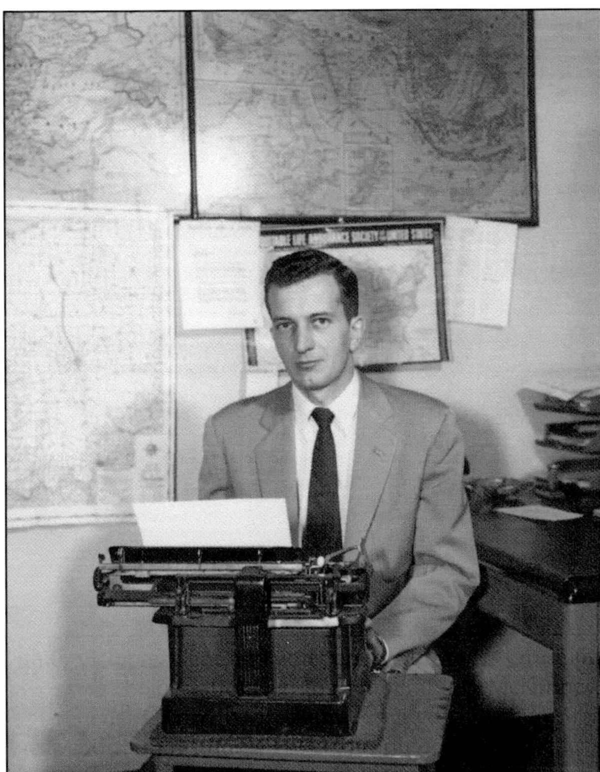

The first news anchor for WISH TV was Vince Leonard. Leonard had been with WISH Radio and made a smooth transition into television. When Bob Lemon, former manager of WTTV, moved to WRCV TV in Philadelphia, he recommended that the company hire Leonard. Leonard moved to Philadelphia in 1958 and was a longtime top-ratings getter there when he "retired" in 1980. He and his wife moved to Phoenix, Arizona, where he finished his news career at KPNX TV.

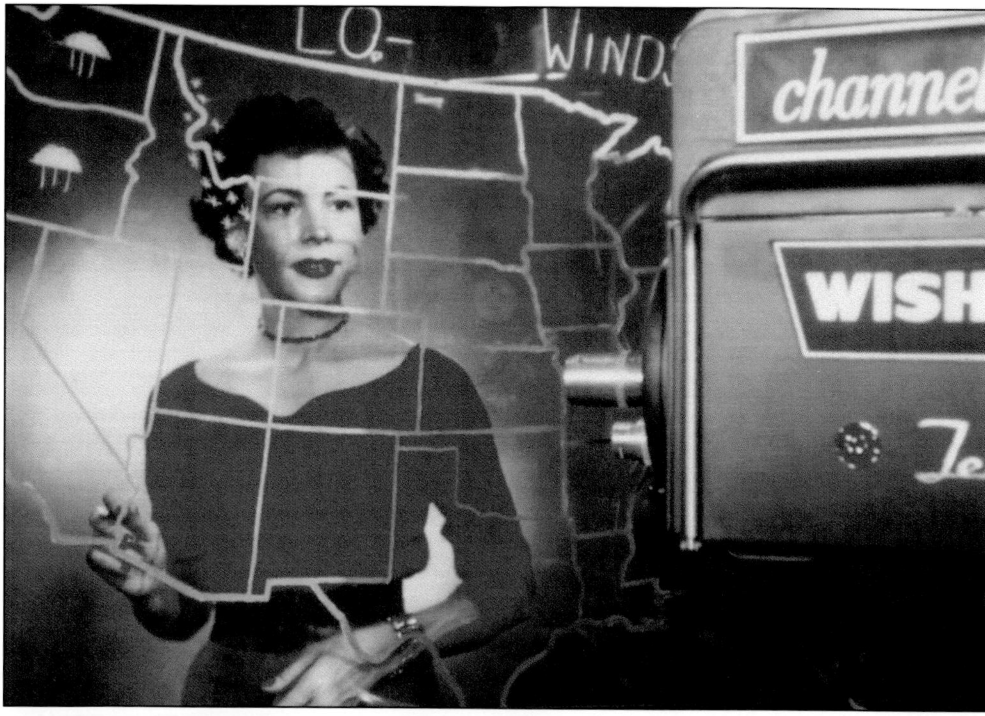

Phyllis Cline was the first weather girl at WISH TV, but she left after just one month. WISH executives found a successor to Cline at WFBM TV. Kay Field was a former band singer and was doing some freelance singing work at channel 6. She auditioned for the weather, was hired, learned how to write backwards, and stepped behind the glass map. Field did the weather for WISH TV from 1954 to 1967.

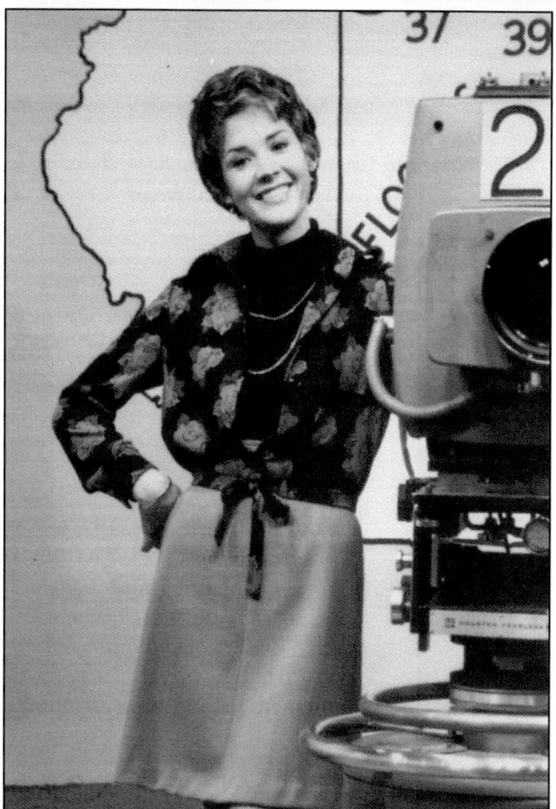

Like mother, like daughter: Kit Field Kruger became a weather girl at WISH TV in the mid-1970s. Kruger also was a host for *The Money Movie* and *Dialing for Dollars* on WISH TV.

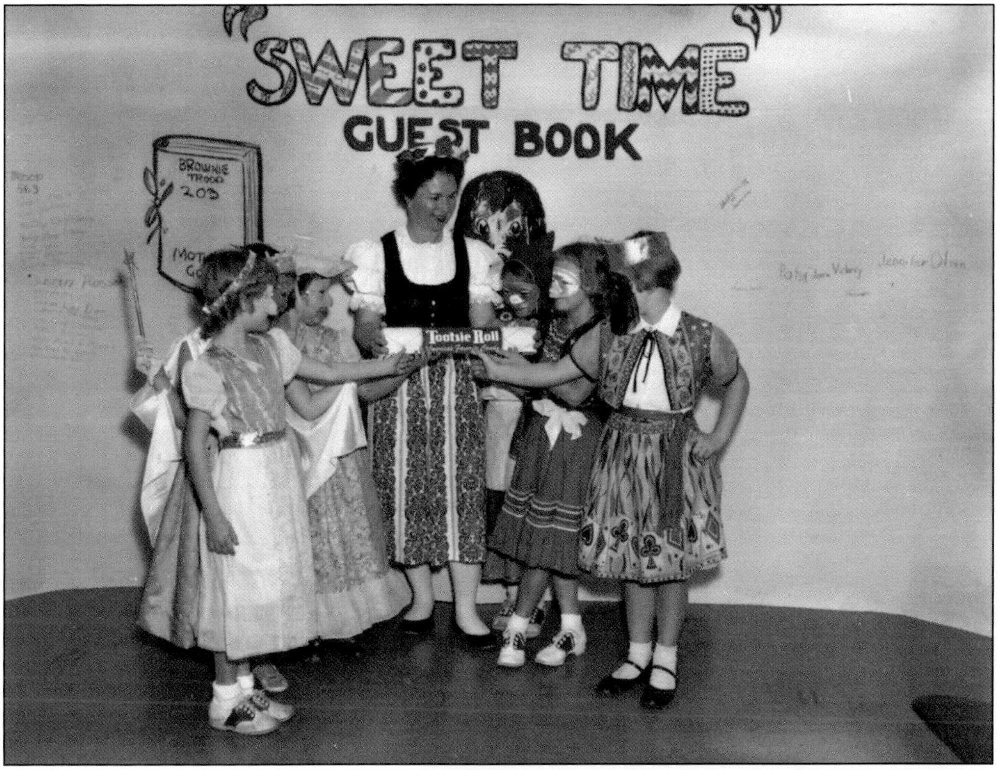

Dee Sweet, aided by her artist-husband, Herb, was host of the first live children's show on WISH TV. The Sweets started the first day camp in America in 1933 and had 10,000 campers over five decades at their Acorn Farm Camp. Herb Sweet's crafts newspaper column was nationally syndicated. The show featured craft-making and live animals, such as bears and snakes.

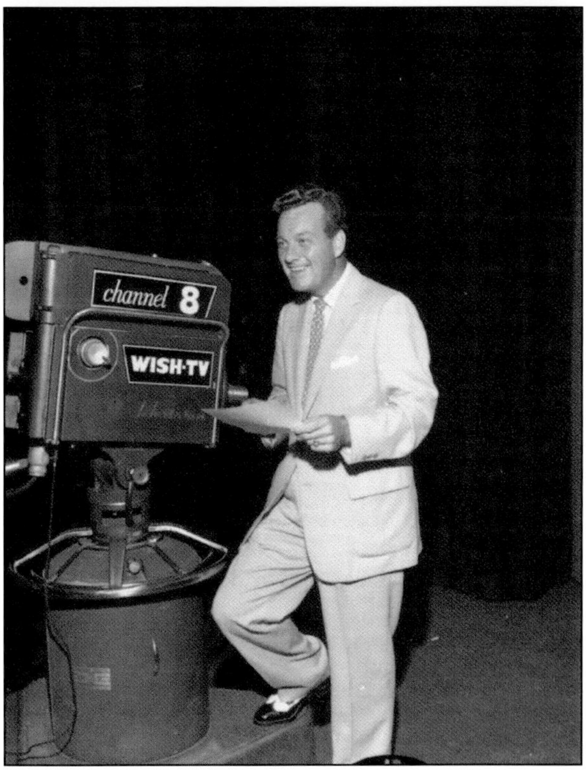

Gene Allison moved from WISH Radio to WISH TV with the greatest of ease and quickly became the most popular and busiest announcer on WISH TV. In the early days, when many commercials were live, including those that were national, Allison was in great demand because of his smooth style, wonderful personality, and pleasing voice.

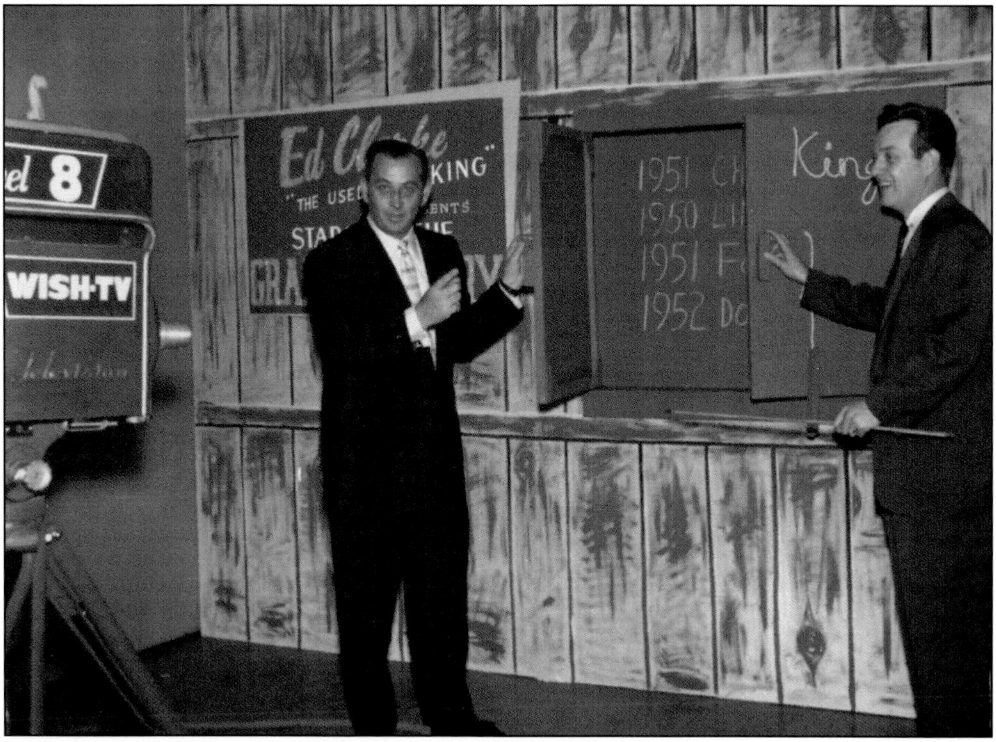

Gene Allison (left) teamed with "Used Car King" Ed Clark to create one of the most colorful and popular duos in the history of early television. Clark bought sponsorship of the entire Saturday *Late Show,* which ran movies, on channel 8. During commercial breaks, Clark would ask Allison to "schneid" (cut) the prices on a list of cars. When Allison started crossing out the prices and writing in lower ones, Clark would exclaim, "Gene, you're killing me!"

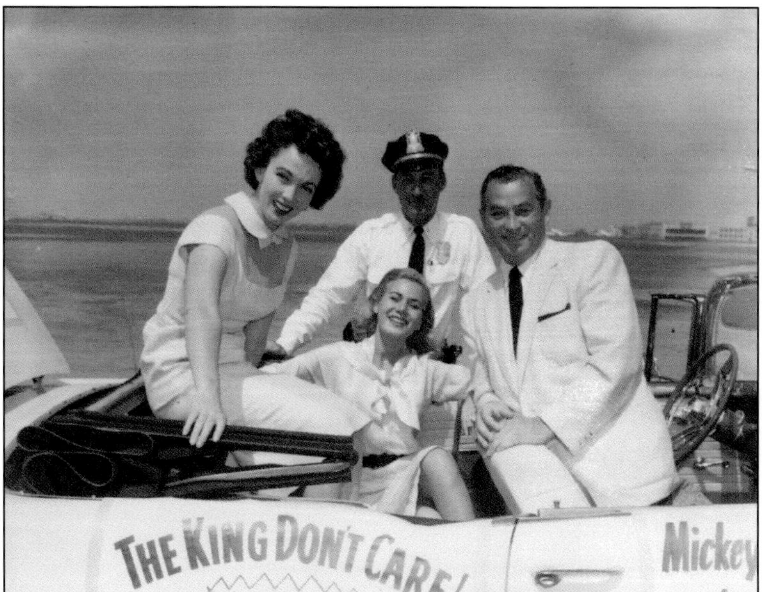

Ed Clark (right) is seen with starlets from the Mickey Spillane movie *Kiss Me Deadly.* For many years, Clark was the biggest advertiser on WISH TV. He told viewers, "Just bring me an honest face," concluding with "the King don't care!"

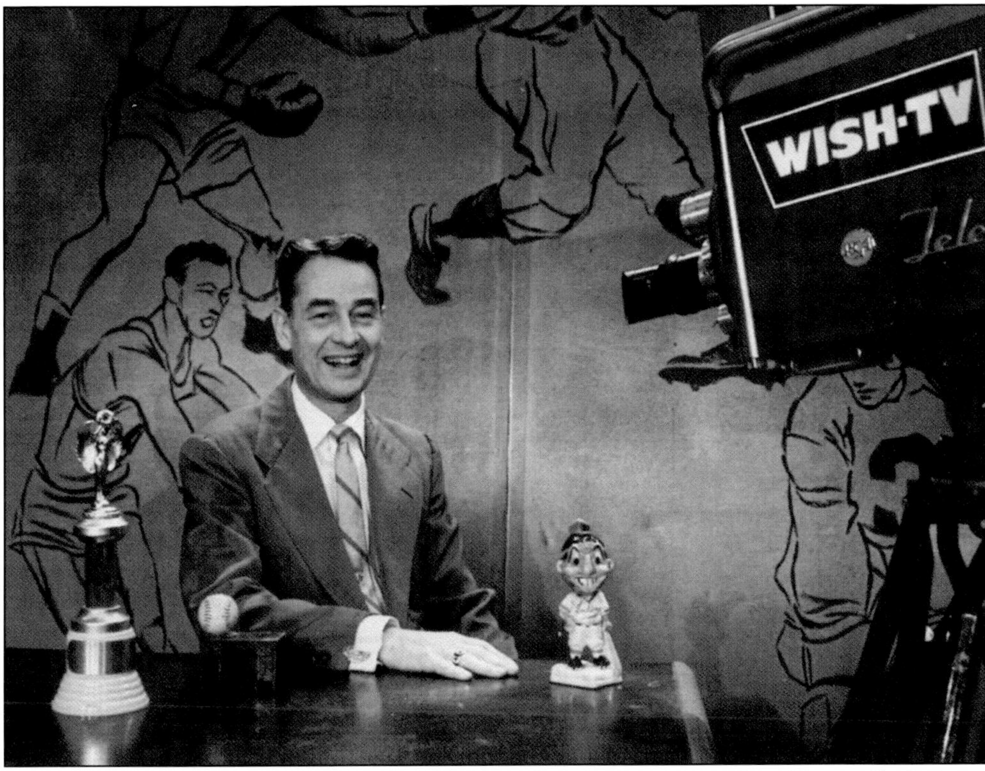

Luke "300-words-a-minute" Walton started his broadcasting career in 1931 in Evansville, Indiana. In 1941, he was hired as a sports announcer/salesman at newly formed WISH Radio. He drew huge audiences for his radio coverage of the Indianapolis Indians baseball team. One of his favorite expressions was, "And there it goes out into Stokely's (the sponsor) pea patch for a home run!" He was the sports director of both WISH Radio and WISH TV.

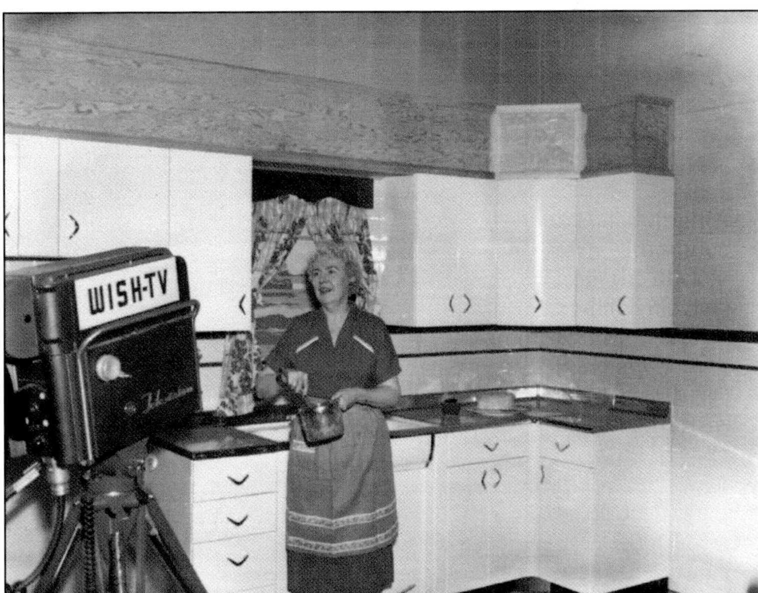

A veteran of 20 years in radio as a home economist, Catherine Daniels made an easy transition from radio to television. Her cooking show *At Home with Catherine Daniels* on WISH TV aired daily for a half hour at 3:00 p.m. A fully equipped kitchen provided some wonderful meals for the staff and technical crew.

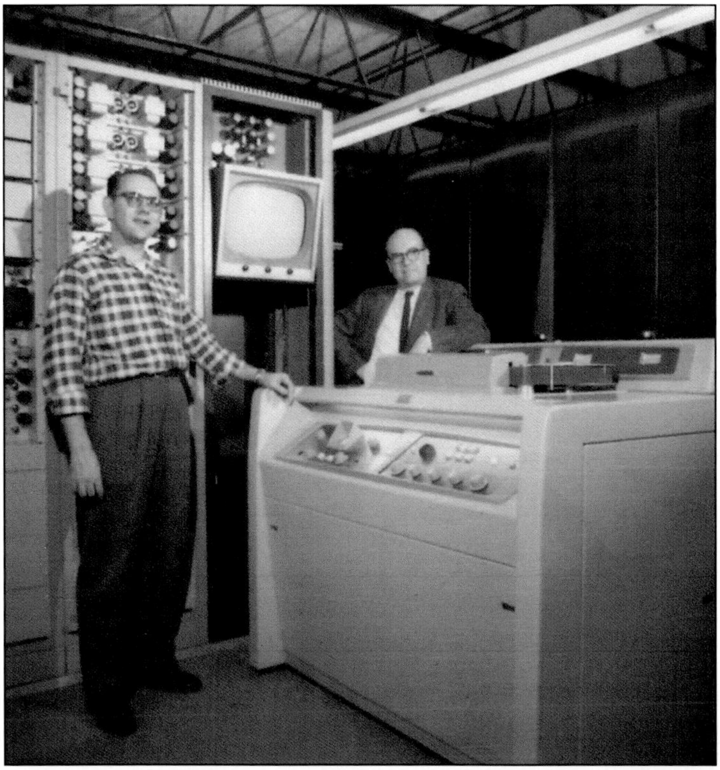

WISH TV owner Bruce McConnell attended the National Association of Broadcasters convention in Chicago in 1956 and saw the first demonstration of a videotape machine—he bought it on the spot. Al Szaley (left) and Bob Brockway were responsible for installing the first videotape machine in Indiana and one of the first in the nation.

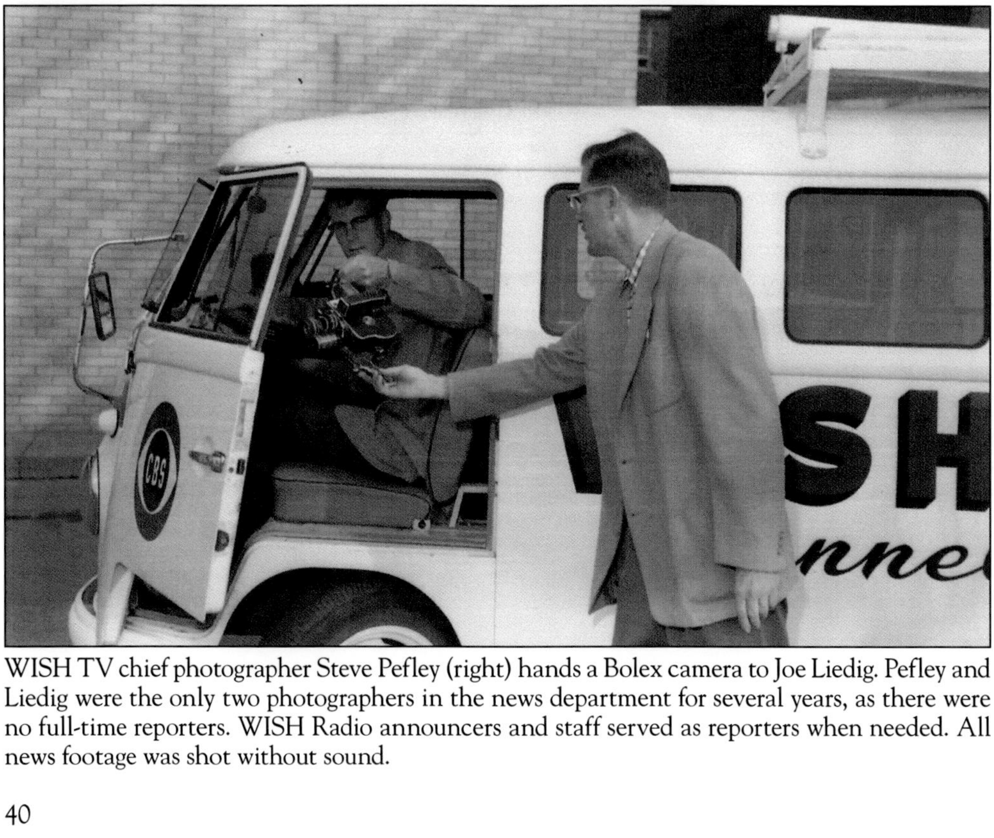

WISH TV chief photographer Steve Pefley (right) hands a Bolex camera to Joe Liedig. Pefley and Liedig were the only two photographers in the news department for several years, as there were no full-time reporters. WISH Radio announcers and staff served as reporters when needed. All news footage was shot without sound.

In the days before videotape, most programming other than network was on 16-milimeter film. Bob Smith was one of three full-time film editors who edited movies, syndicated shows, and news film. Smith later became a director and then production manager at WISH TV.

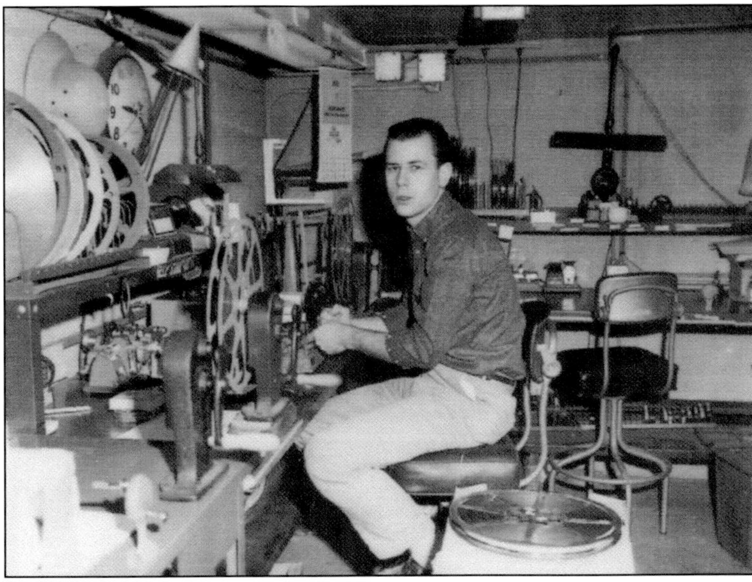

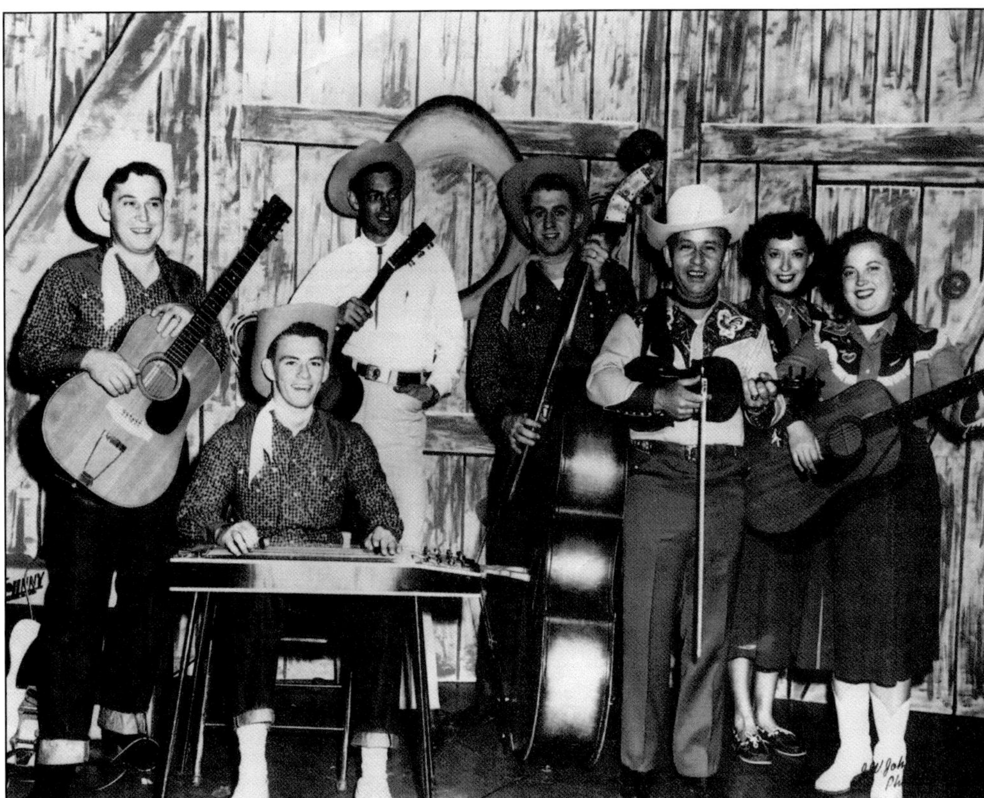

Shorty Sheehan (third from the right) headed a group of country music performers in a program called *Hillbilly Heaven*. Sheehan brought his show to WISH TV shortly after the station signed on in 1954. His program was fully sponsored by Abel's Auto and aired every Saturday.

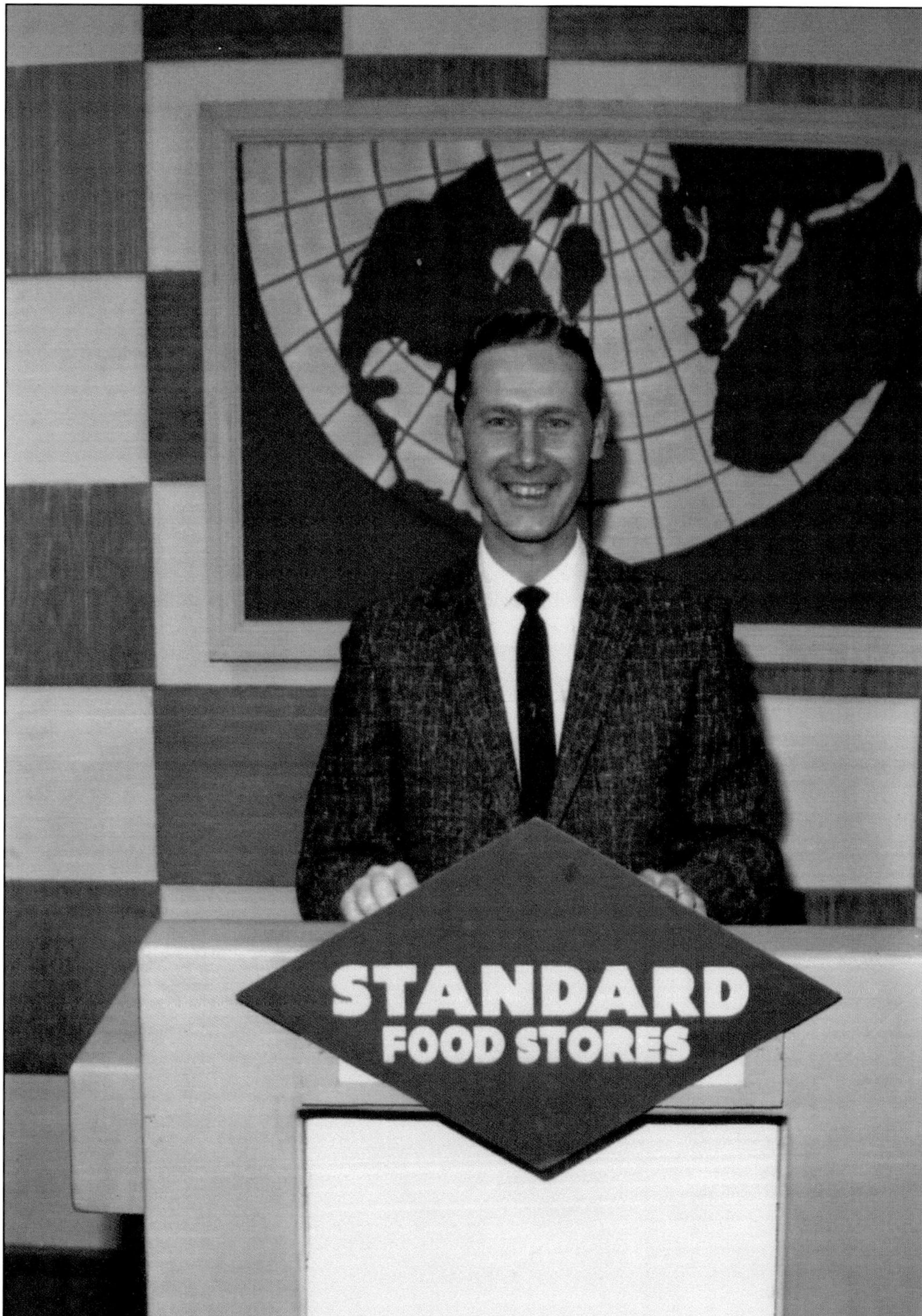

Bill Donnella was the anchorperson for the daily 1:00 p.m. news on WISH TV. He joined WISH TV in 1959. In addition to news, he did sports, including Indianapolis Indians baseball games and the Indianapolis 500 Mile Race. He joined WIFE Radio in 1965 as news director. He also was associated with WTHR TV, WIBC, and WXIR.

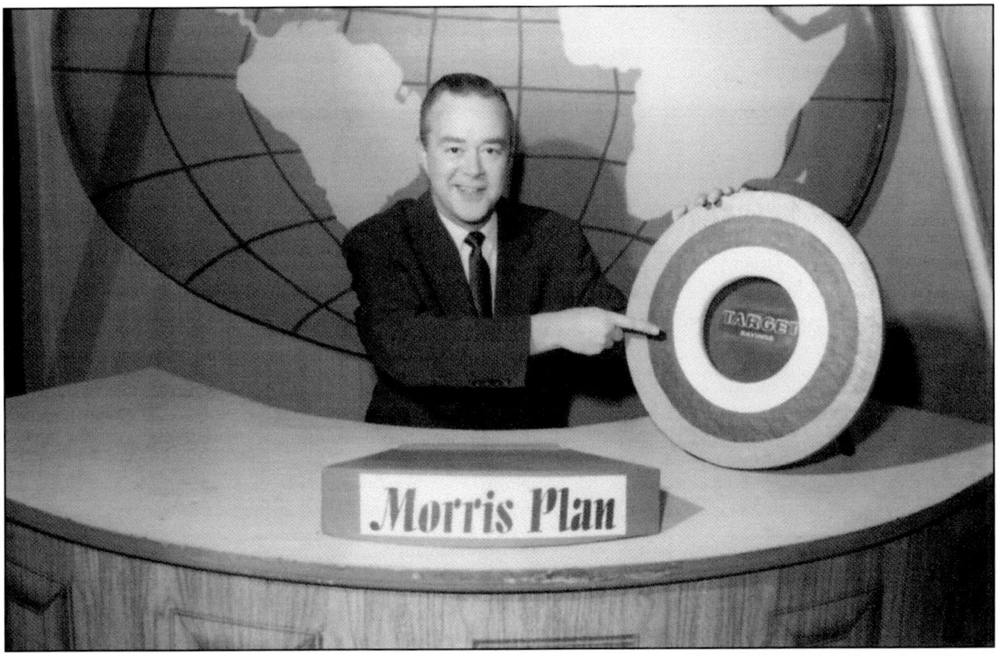

In 1958, when Vince Leonard left WISH TV, Richard Hickox replaced him. Hickox had been the anchorperson at WANE TV in Fort Wayne, which was WISH TV's sister station. Note the Morris Plan logo on the desk. Indianapolis Morris Plan had full sponsorship of the 11:00 p.m. news five days a week.

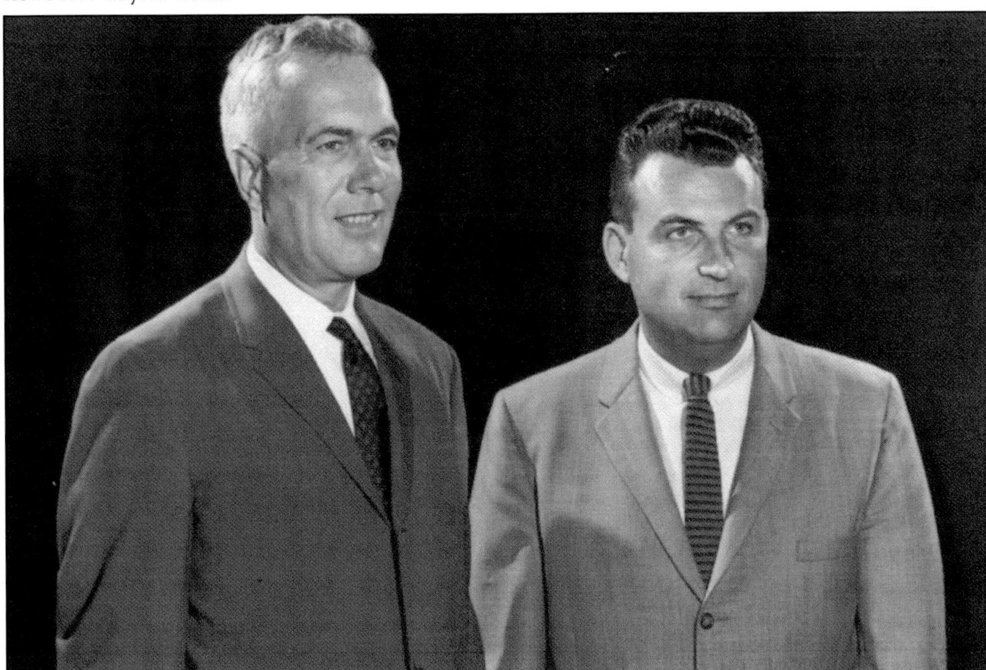

In 1951, Howard Caldwell (right) left WTHI TV in Terre Haute to become the primary anchorperson for WFBM TV. Caldwell and weatherman Bill Crawford continued to dominate the news ratings in the Indianapolis market.

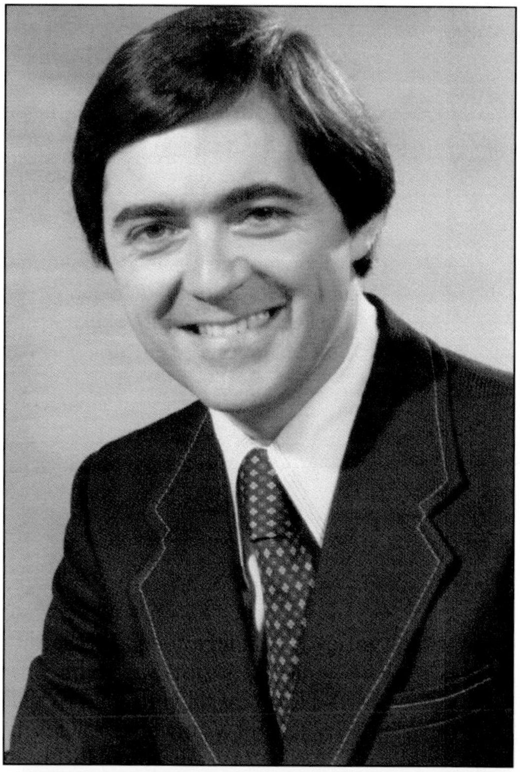

When Mike Ahern joined the WISH TV news department in 1967, he became a coanchor. He was first placed with Bill Aylward, then with Bob Glenn, and, finally, with Rolland Smith. When Glenn, Aylward, and Smith successively left, Ahern became the sole anchor. In 1980, Debby Knox joined him as coanchor. Ahern retired in 2004 after more than 30 years as a news anchor.

After signing on the air in 1957, WLWI TV (channel 13) began to put its news department together. Gordon Graham (seated) was named anchorperson and news director. Forest Boyd (to the right of Graham) was an anchor/reporter, and Bill Landrum (far right) was the chief photographer.

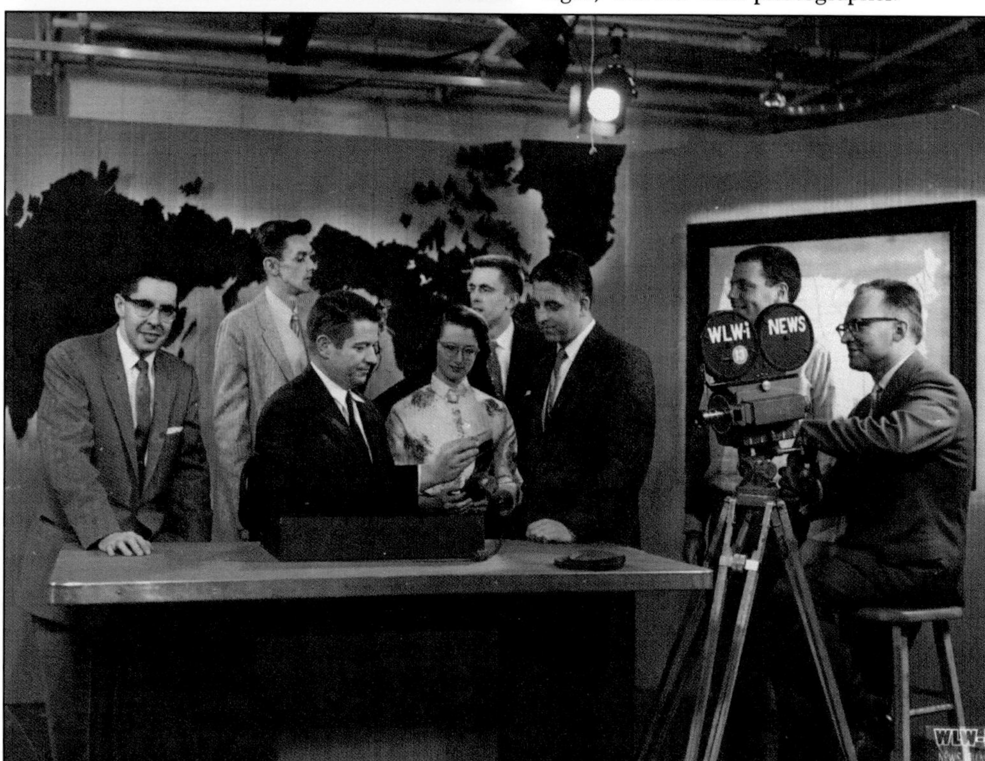

In 1959, WFBM TV hired Bob Gamble as news director to replace Ken Mayer, who had died suddenly. Gamble came from WKY TV in Oklahoma City, where the Radio and Television News Director's Association had named his station "Outstanding TV News Operation." Gamble immediately began strengthening the news department and was primarily responsible for its continued leadership throughout his tenure at the station.

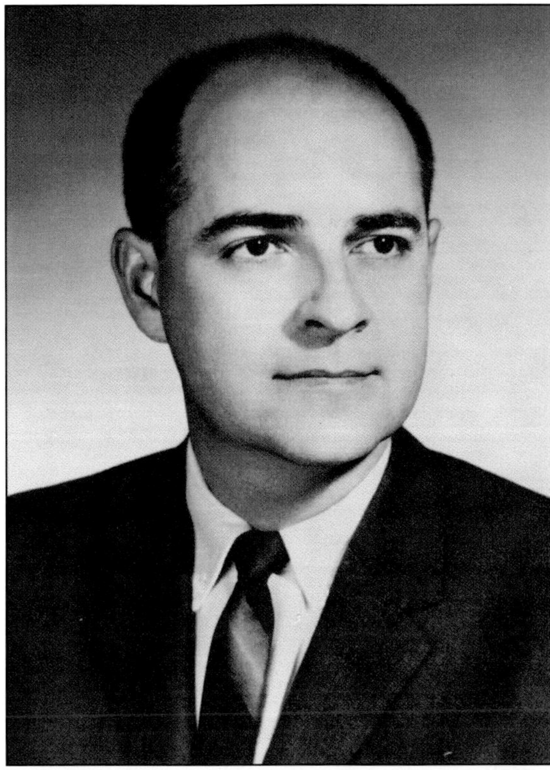

Three news photographers covering the same news story are, from left to right, Ray Haas, WISH TV; Dick Baldwin, WFBM TV; and Bill Landrum, WLWI TV.

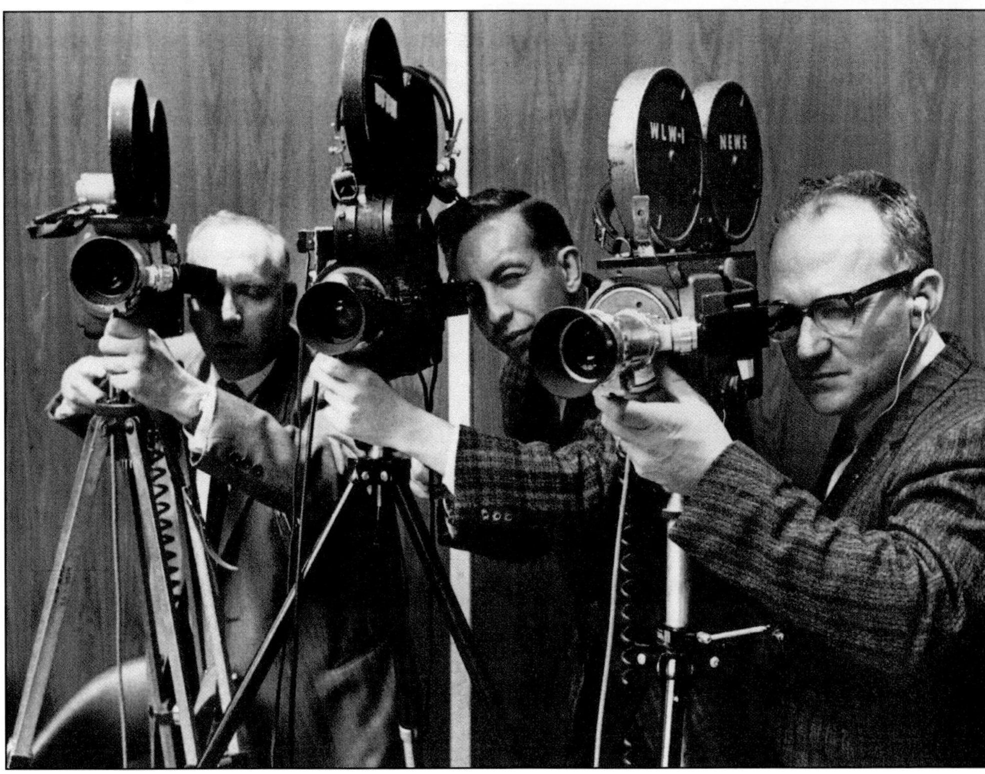

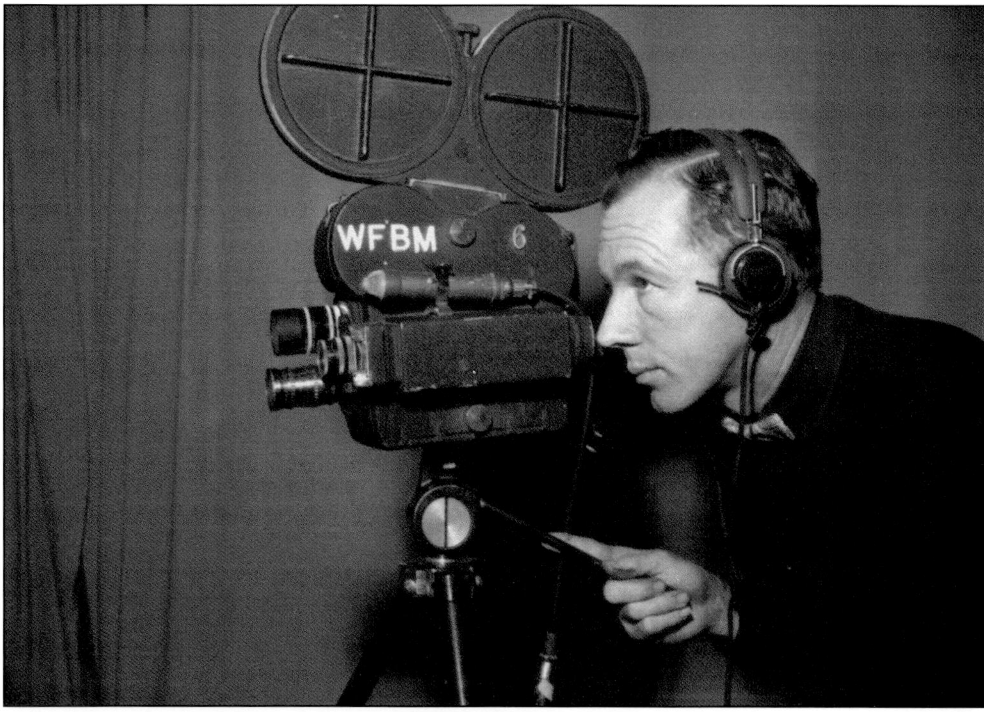

Ernie Crisp was chief photographer at WFBM TV from 1959 to 1970. He was named "Television News Photographer of the Year" in 1966. He also served a term as president of the Indiana News Photographers Association. He left WFBM TV in 1970 to join Eastman Kodak. He was killed in 1977 at the age of 49 when the 1941 Waco biplane he was piloting crashed near Rochester, New York.

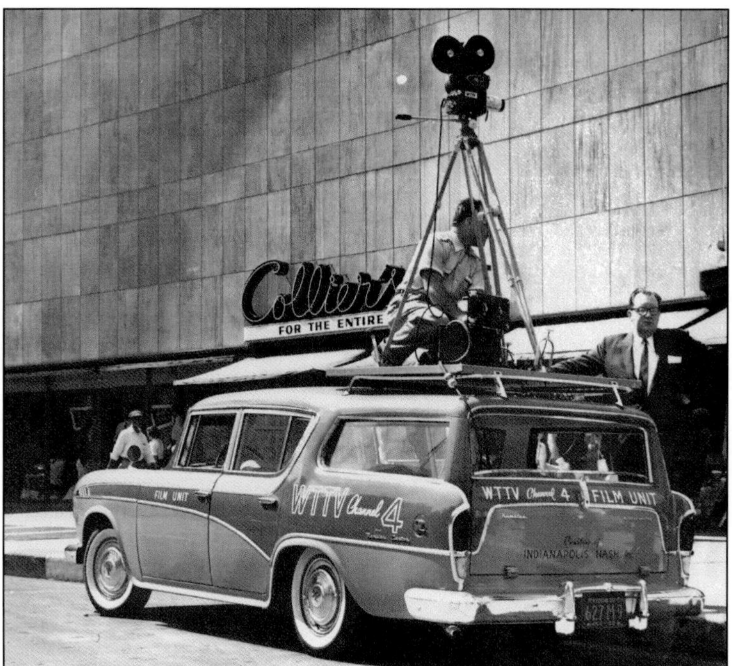

Frank Edwards (right) is shown on Monument Circle in 1959 with photographer Bill Landrum. As news director for WTTV, he oversaw a very slim news department. Shortly after World War II, Edwards was hired by the Mutual Broadcasting System to do a nightly newscast sponsored by the American Federation of Labor. He left the network in 1954. In 1955, he joined WTTV as a news commentator. In 1965, he made a brief move to WLWI TV.

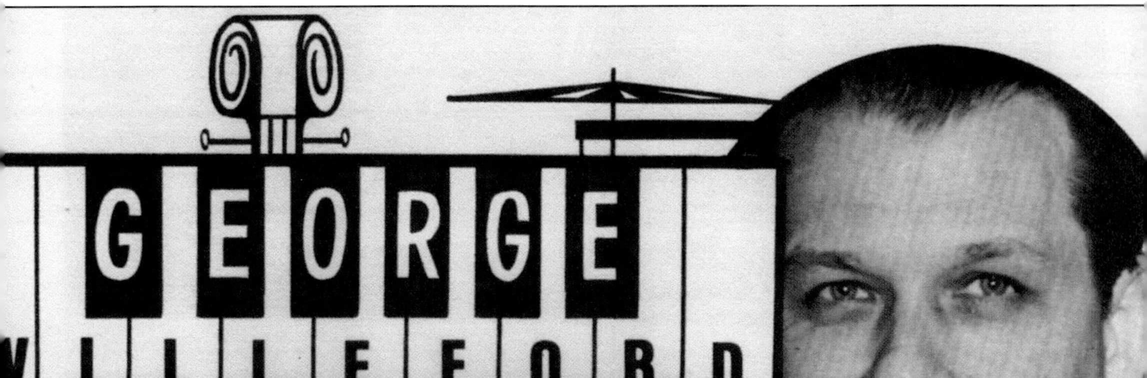

George Willeford began hosting a talk/variety show in 1960 on WLWI TV. Murray Horn, a baritone singer, served as a fall guy for Willeford, while the Don Wilhite Trio provided music. The show was a mixture of wry humor, music, and information. It had a relatively short run. Willeford had previously been a host of *Movie House*, in which he made fun of the movies he was showing. He also occasionally did the weather.

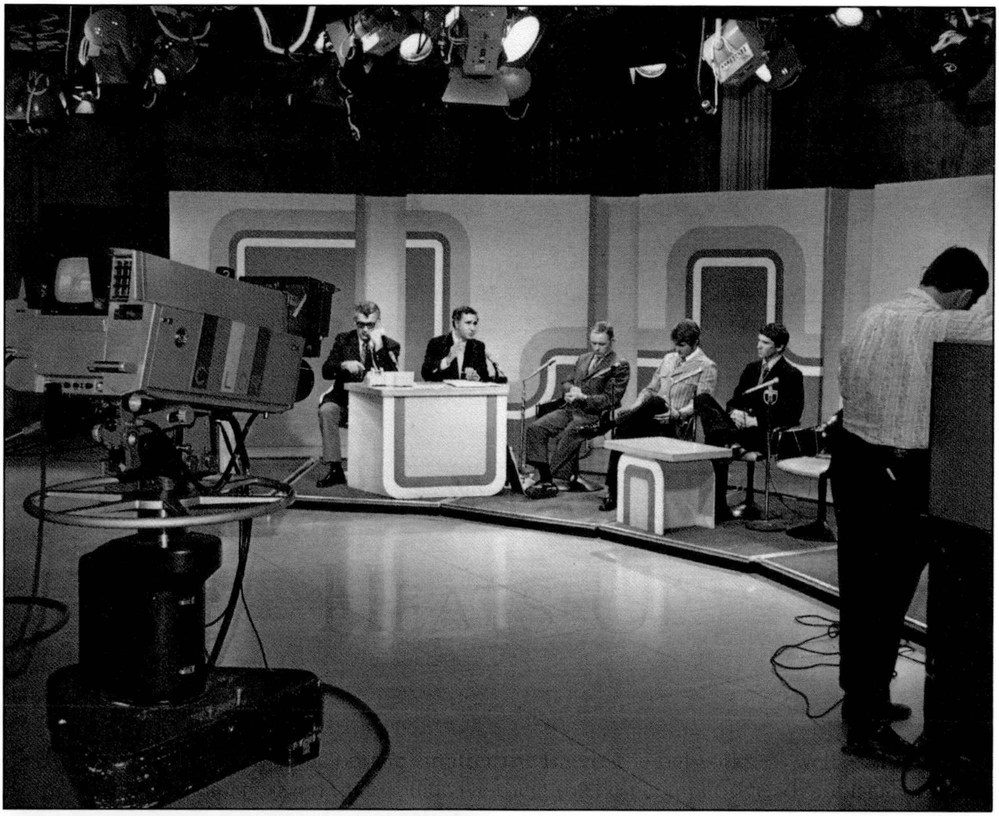

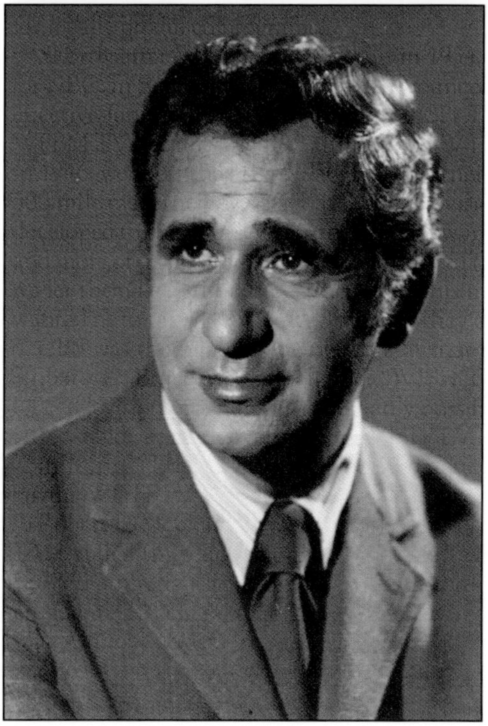

The Jim Gerard Show began on WFBM TV in 1962 and was on daily almost continuously from 1962 to 1987. Anyone who was anybody came to *The Jim Gerard Show*.

The show ended on WFBM TV when Gerard left to join Hook's Drug Company. Shortly thereafter, Gerard resumed the show on WTTV with Hook's as a sponsor. The show ended when Revco absorbed Hook's in 1987.

Six

THE COMPETITION HEATS UP

As TV stations began to come on the air after the freeze, the networks, as well as the stations, started jockeying for a position that would give them the best competitive advantage. For almost four years, WFBM TV aired programs from all four networks and freely preempted network shows for locally produced programs.

Shortly after WISH TV signed on in 1954, it dropped ABC and became the exclusive outlet for NBC. In September 1954, it was announced that WFBM TV and CBS were parting company effective June 2, 1955. At that time, WISH TV gave up NBC to become a CBS affiliate, while WFBM TV became an NBC affiliate and WTTV with ABC. DuMont, which began its network in 1946, went off the air in 1956. Harry Bitner Jr. explained, "When we went on the air with the 500 mile race, there were about 2,000 television sets tuned in. As time went on and the market developed to the 600,000 sets there are now in central Indiana, Columbia wanted us to use more and more of its programs. We felt a loyalty to those who had believed in us in the beginning, and we let them keep their time schedules. This meant we could not take as many programs as CBS wanted us to take. Maybe they thought they could get more time on a newer station."

CBS was the number-one network at that time, and WISH TV was delighted to become its affiliate. C. Bruce McConnell said, "WISH TV intends to carry as many of the CBS radio and CBS television programs as is humanly possible, and it will carry them 'live.' " This meant that CBS would get more exposure on WISH TV.

In 1956, C. Bruce McConnell received notice that the J.H. Whitney Company was interested in buying WISH TV, as well as WANE TV in Fort Wayne, and associated radio stations. According to Robert B. McConnell, a broker asked his father how much he wanted for the stations. He said, "Dad pulled this out of no place and said, '$10 Million.' " Two weeks later, Whitney agreed to pay $10 million; at that time, it was a record price. Robert B. McConnell was retained as vice president and general manager and stayed in that position until 1979.

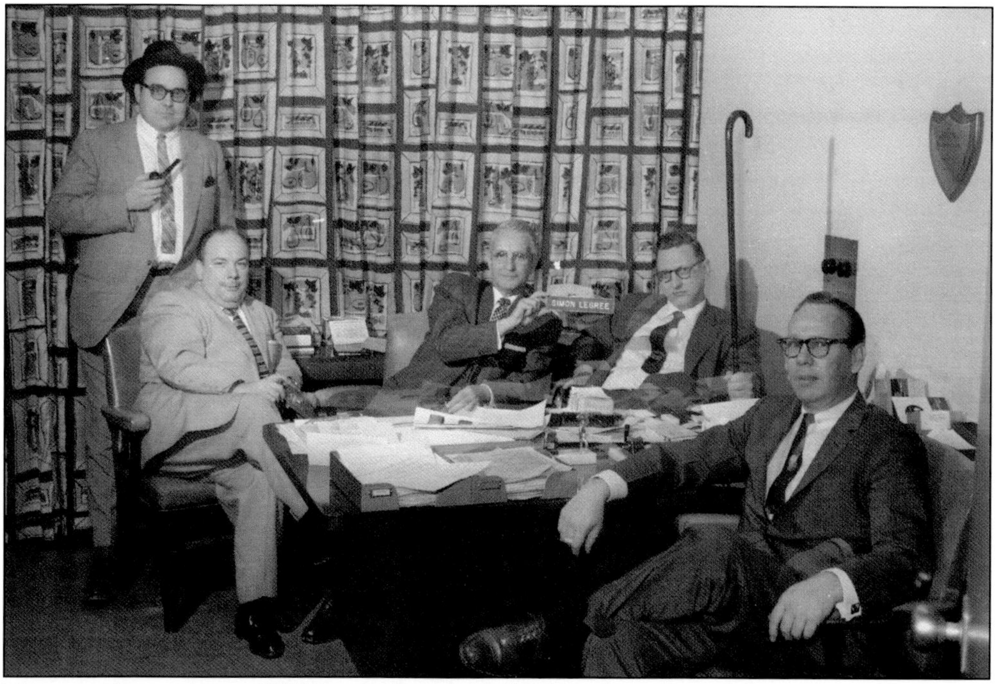

Here is a sales meeting at WISH in 1957 in the Riddick Building. From left to right are Bill Eagen, Tommy Moriarity, Leonard Bishop, Bob Ohleyer, and Bill Faulkner. Ohleyer was the WISH Radio sales manager, and he took on the additional responsibilities of WISH TV sales manager for the first year or two after it signed on.

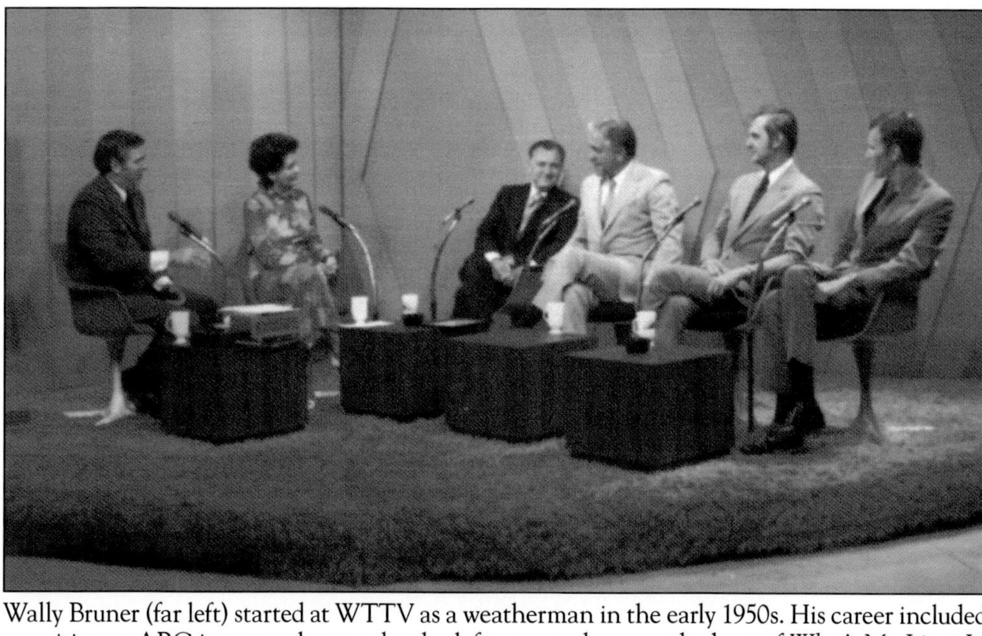

Wally Bruner (far left) started at WTTV as a weatherman in the early 1950s. His career included a position at ABC in network news, but he left news to become the host of *What's My Line?* In the late 1960s, Bruner came back to Indiana and, with his wife, Natalie, hosted *The Morning Show*, a daily one-hour talk show on WISH TV. Unfortunately, the show had a short run, and Wally turned to producing and syndicating his *Wally's Workshop* series.

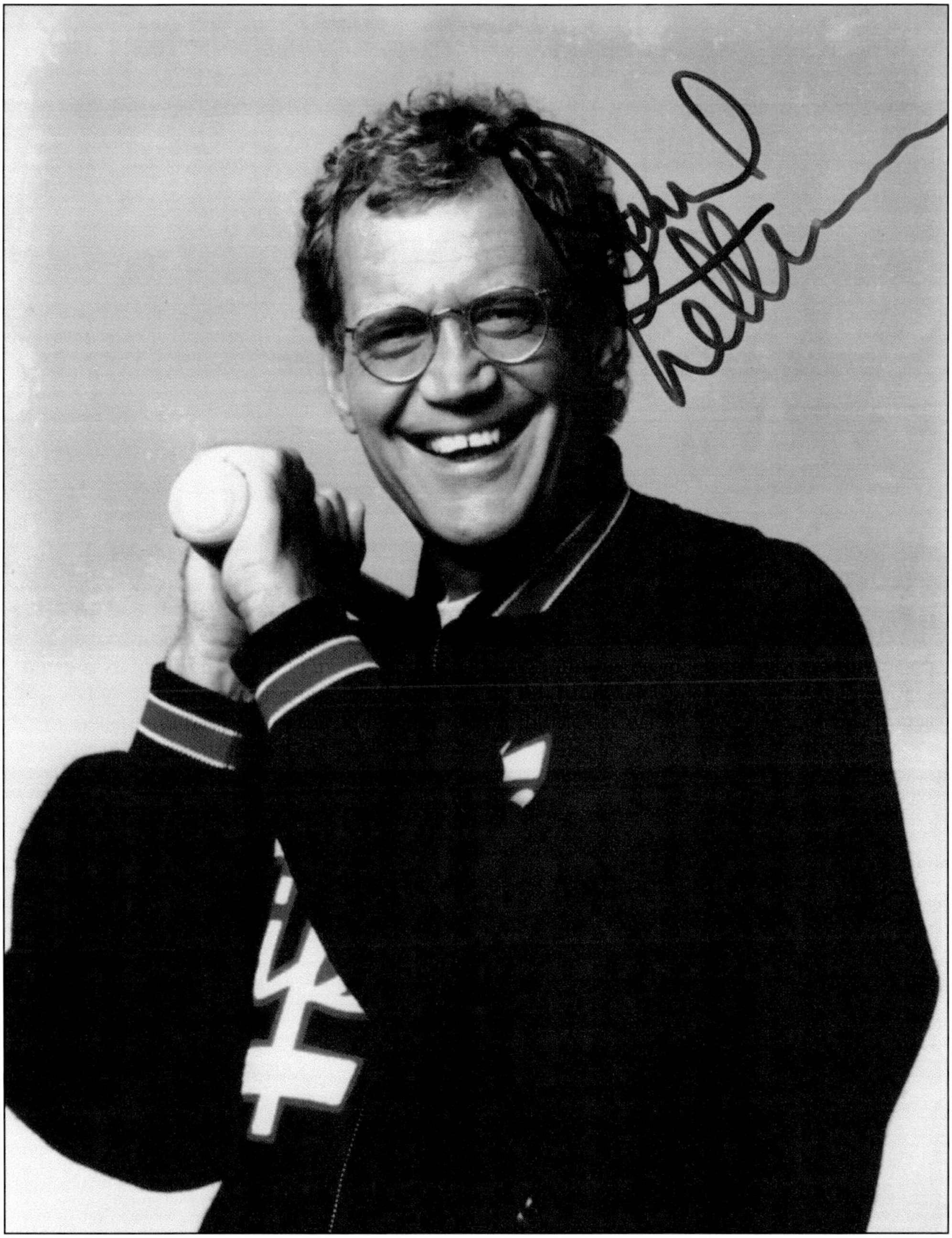

In 1969, David Letterman graduated from Ball State University in Muncie, Indiana, and joined WLWI TV in Indianapolis. He became a wacky weatherman, making up cities and erasing borders between states. He hosted a movie series, *Freeze Dried Movies*, and a 4-H program called *Clover Power*. He left WLWI TV in 1974 to host his own talk show on WNTS in Indianapolis. In 1975, he took off for Los Angeles.

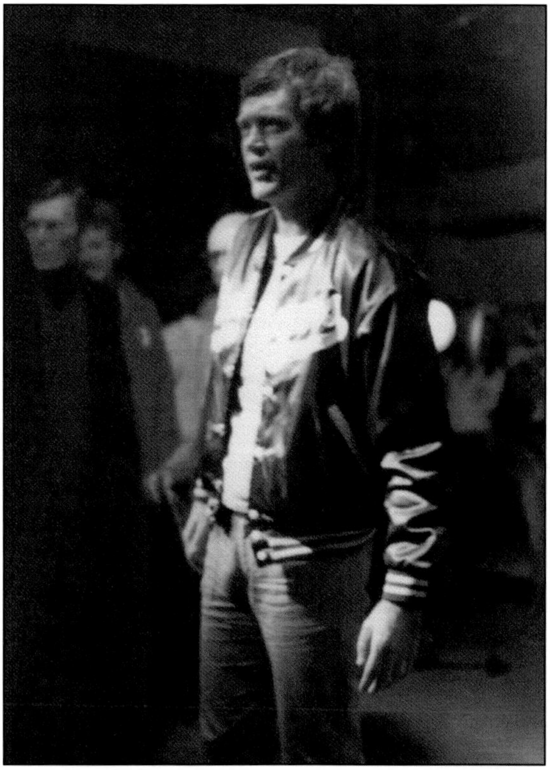

Letterman came back to Ball State in 1979 as the homecoming attraction. From the stage, he announced, "I'd like to thank Ball State for giving me back my entire four-year tuition in one night."

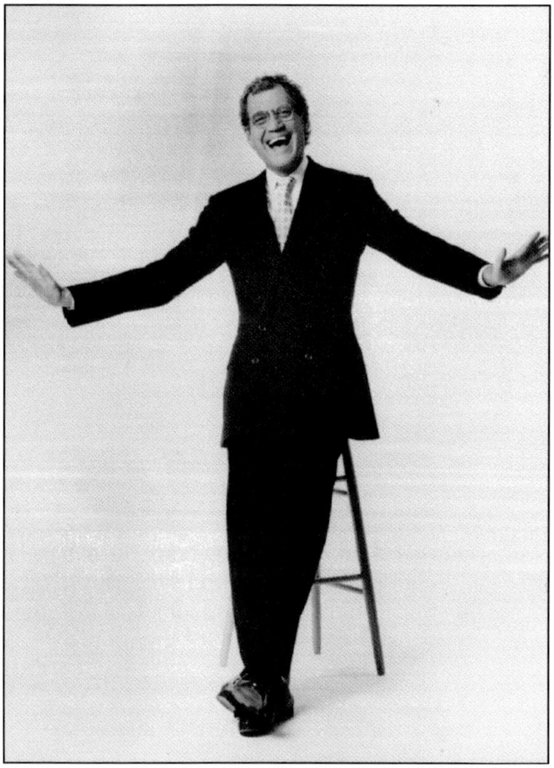

Letterman left channel 13 in Indianapolis to move to a small local radio station. Not having to appear on television, he grew a rather heavy beard. WNTS Radio gave Letterman an afternoon talk show. He stayed in that position for less than a year and headed for Los Angeles.

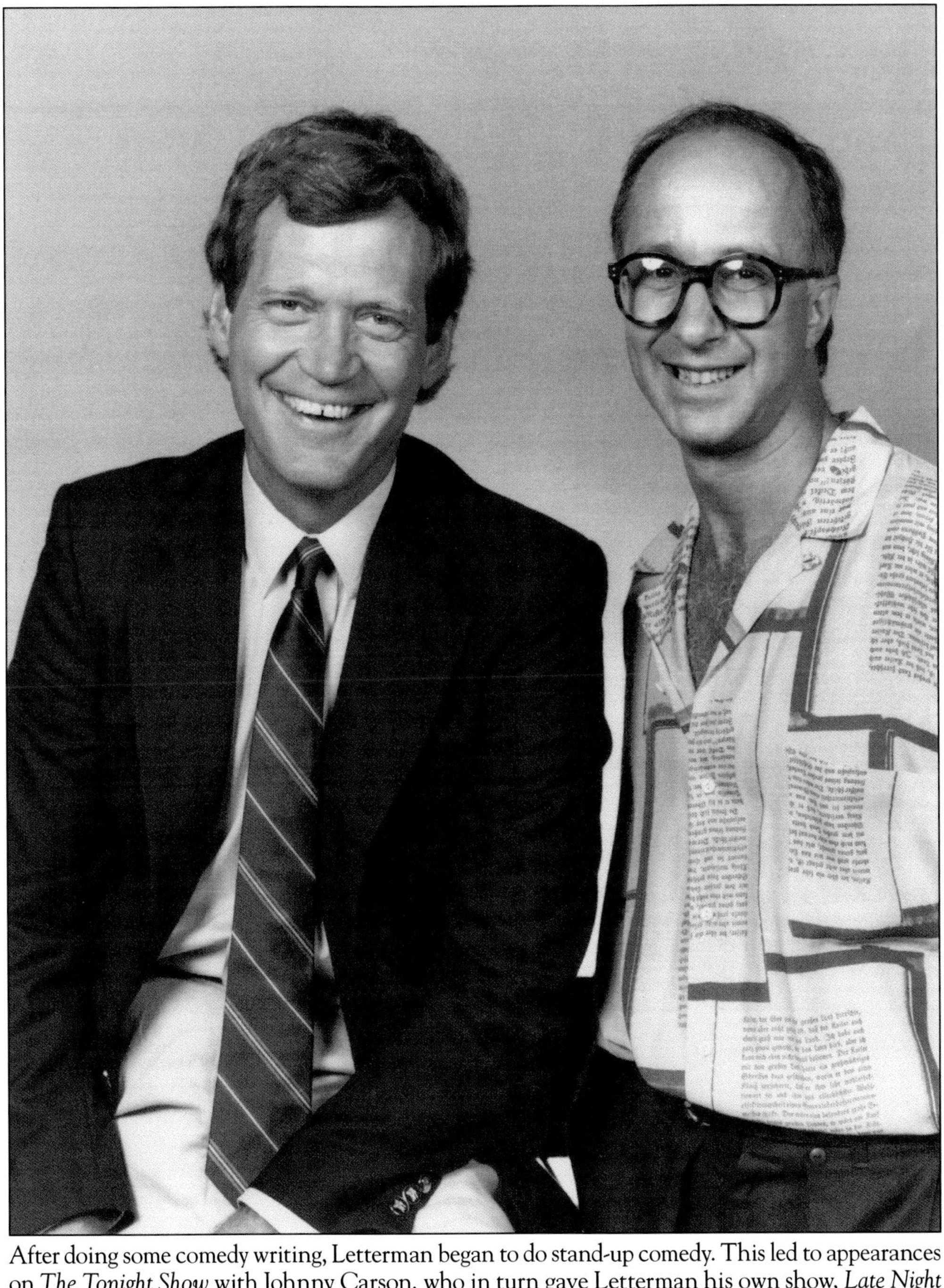

After doing some comedy writing, Letterman began to do stand-up comedy. This led to appearances on *The Tonight Show* with Johnny Carson, who in turn gave Letterman his own show, *Late Night with David Letterman*. Paul Shaffer was the orchestra leader for Letterman's NBC show and moved with him to CBS.

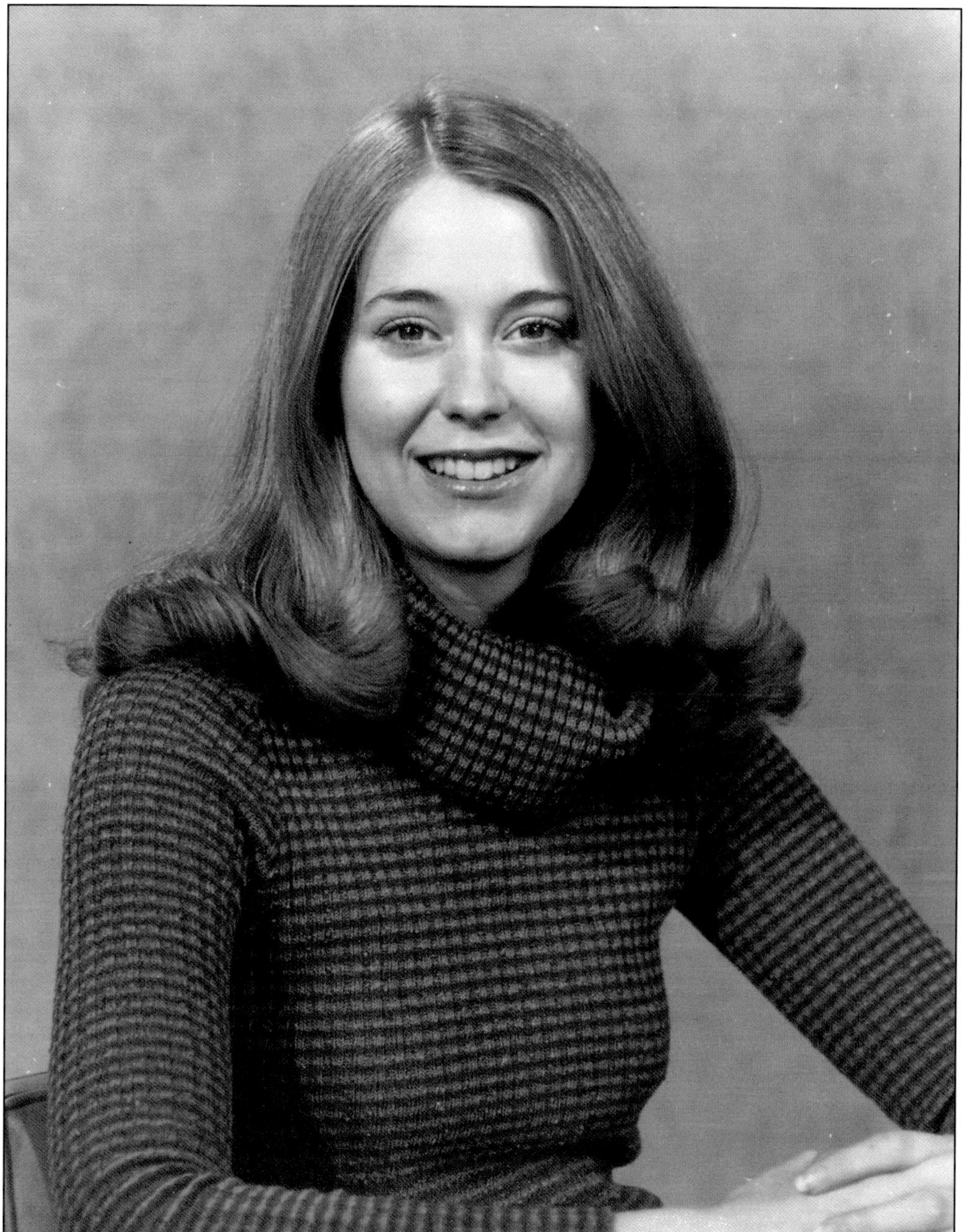

Jane Pauley walked in the door at WISH TV in 1972 and was ushered to news director Lee Giles, who was trying to get more females in the newsroom. Giles auditioned her, but it took him three months to convince the general manager to take a chance on her. She had no experience and no college training. She majored in political science at Indiana University but never took a broadcasting course.

Pauley was hired as a general assignment reporter and was a quick learner. She began anchoring the weekend news and was soon spotted by WMAQ TV (NBC) executives, who were similarly under the gun to increase the number of females in news. She moved to Chicago and became the first female anchor at WMAQ TV. Barbara Walters quit the *Today Show*, and NBC began auditioning replacements. After just four years in television, she was hired to coanchor *Today* with Tom Brokaw.

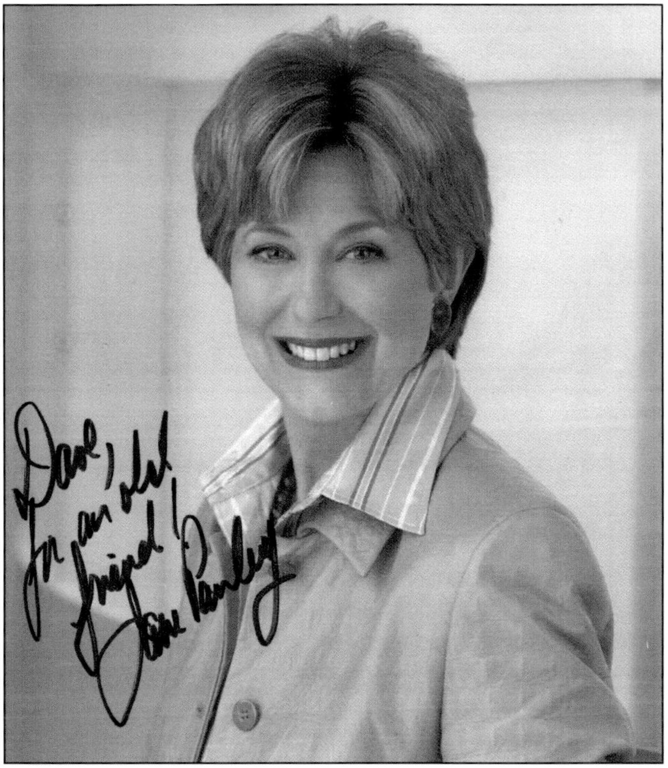

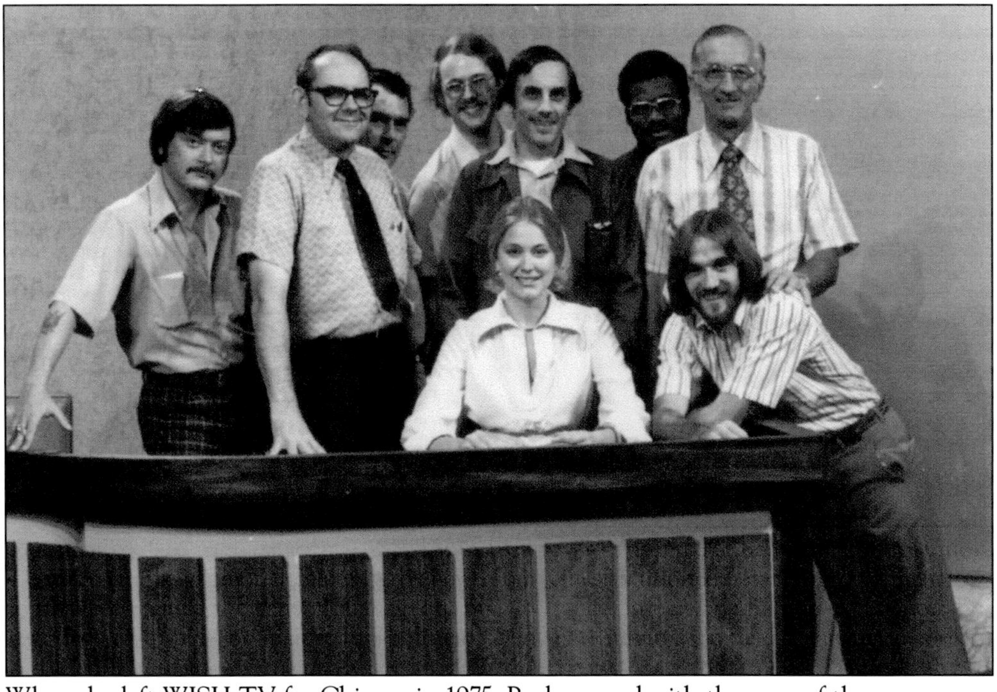

When she left WISH TV for Chicago in 1975, Pauley posed with the crew of the noon news. Surrounding Pauley are, from left to right, Jim Pasierb, Al Szaley, Charlie Clark, Andy Armstrong, Bob Warren, Fred Leggs, Joe Missick, and Kip Sexson.

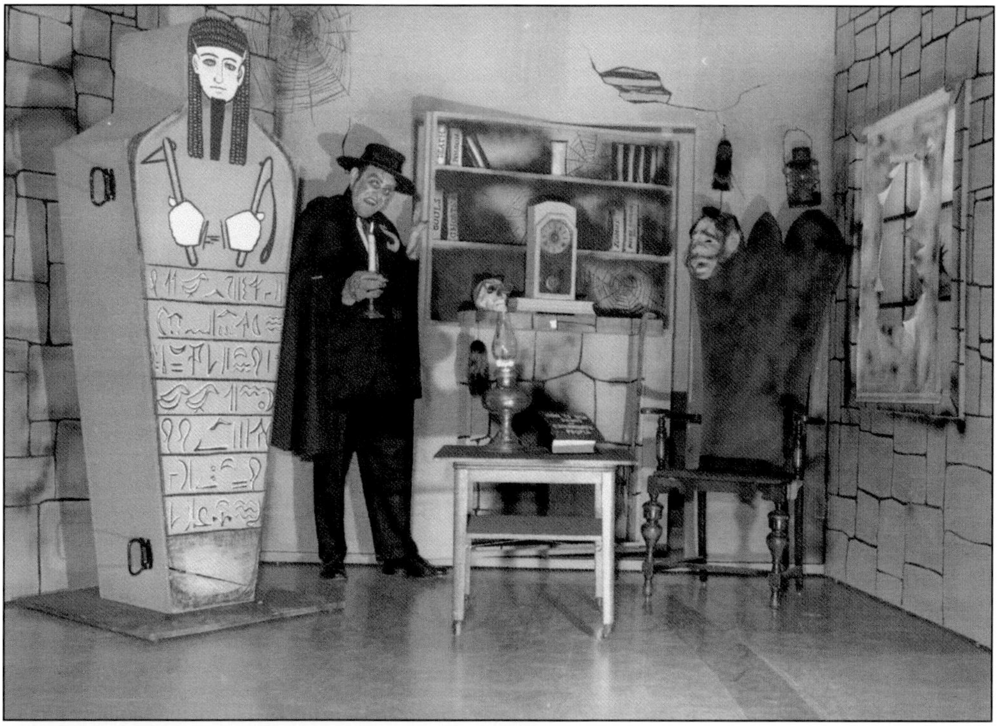

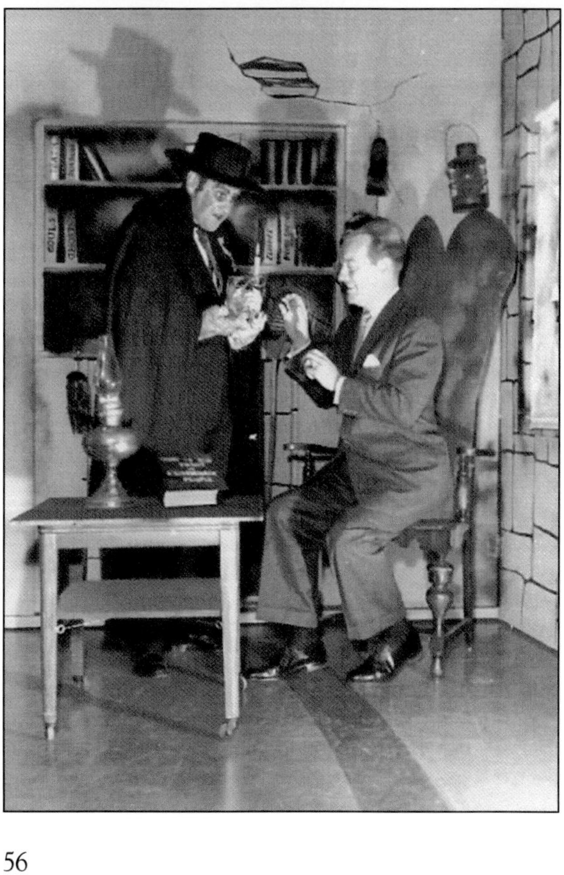

In 1958, WISH TV bought the Universal Studios package of horror films. This meant that viewers in the Indianapolis television area would be seeing films like *Dracula*, *Frankenstein*, *The Wolf Man*, and other horror classics for the first time on television. A host named "Selwin" was created with WISH TV producer/director Ray Sparenberg playing the character. A set was built enabling Selwin to enter each night through a creaking, revolving bookcase.

Selwin and the Universal movies attracted huge audiences. WISH TV bought a second package of Universal horror films, and Selwin appeared every Friday night until 1961. When the horror films ran out, WISH TV purchased a package of Tarzan and Bomba the Jungle Boy films, and *Selwin on Saturday* became a popular program. He wore a pith helmet and carried his trusty "Selwinchester."

Seven

The End of Local Ownership

The principal of the J.H. Whitney Company was "Jock" Whitney. In the 1950s, Whitney bought five television stations under the banner of Corinthian Broadcasting. Whitney was the brother-in-law of William Paley, the CEO of CBS. Because of this, Whitney bought only stations that were CBS affiliates.

Corinthian owned WISH TV from 1956 to 1984, when it, as well as its companion station in Fort Wayne, was sold to LIN Broadcasting. Local ownership of television stations was quickly disappearing. Crowell-Collier made an offer to purchase WFBM Radio and TV, but the sale fell through, and Time, Inc., offered to buy the Indianapolis radio/television combination as well as other Bitner radio and TV stations. Time, Inc., agreed to pay $15.7 million for all of the Bitner-owned stations. The FCC approved the deal, and Time, Inc., became the owner of the stations on May 23, 1957. In 1971, McGraw-Hill purchased all five of the Time, Inc., television stations for $69.3 million. After this sale, the call letters for the Indianapolis station were changed to WRTV.

The last VHF channel available in Indianapolis was channel 13, and the fight for it was extremely contentious. The four leading contenders for the channel were Crosley Broadcasting (owner of the WLW stations) in Cincinnati, Indianapolis Broadcasting, Inc. (owner of WIRE Radio and Central Newspapers of Indiana), Midwest Television Corp. (headed by Union Federal CEO George Sadlier), and WIBC Radio (owned by Richard Fairbanks). There was a 14-year legal battle for channel 13. The applicants continued to pursue the possibility of buying the channel until September 13, 1962. At that point, the FCC voted four to three to give the channel to Crosley permanently.

At the same time, the FCC approved the sale of WLW-A in Atlanta, Georgia, to WIBC for $2,251,000. Crosley also rewarded WIBC, a leading contender for the channel, by returning all the money invested in pursuing it. Speaking for the three dissenting commissioners, FCC chairman Newton Minow said, "A horse-trade is a perfectly appropriate conclusion to a bargaining session between dealers in horses. But it is not appropriate for a hearing designated to select the most responsible licensee of a publicly franchised television channel."

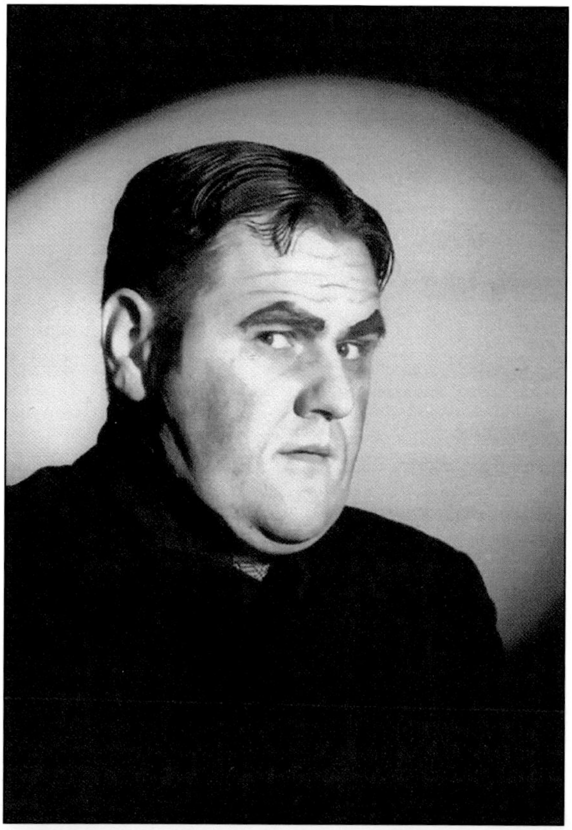

Selwin's last resurrection had him in a space suit on "an unknown planet in outer space." This time, he hosted science-fiction films. His last appearance was in 1963.

In 1962, WTTV purchased the Universal "shock" package that WISH TV had already run. At first, it ran them with no host, but then Bob Carter created "Sammy Terry" as the host. It was very successful, and Carter continued that character, with the occasional hiatus, until 1989. Today, his son makes appearances as Sammy Terry.

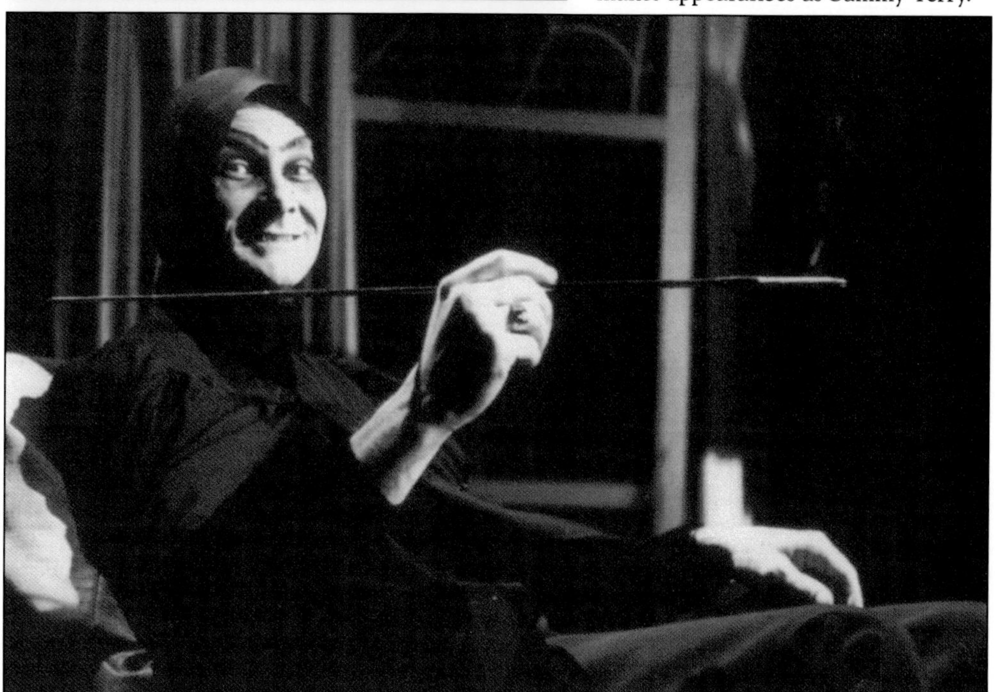

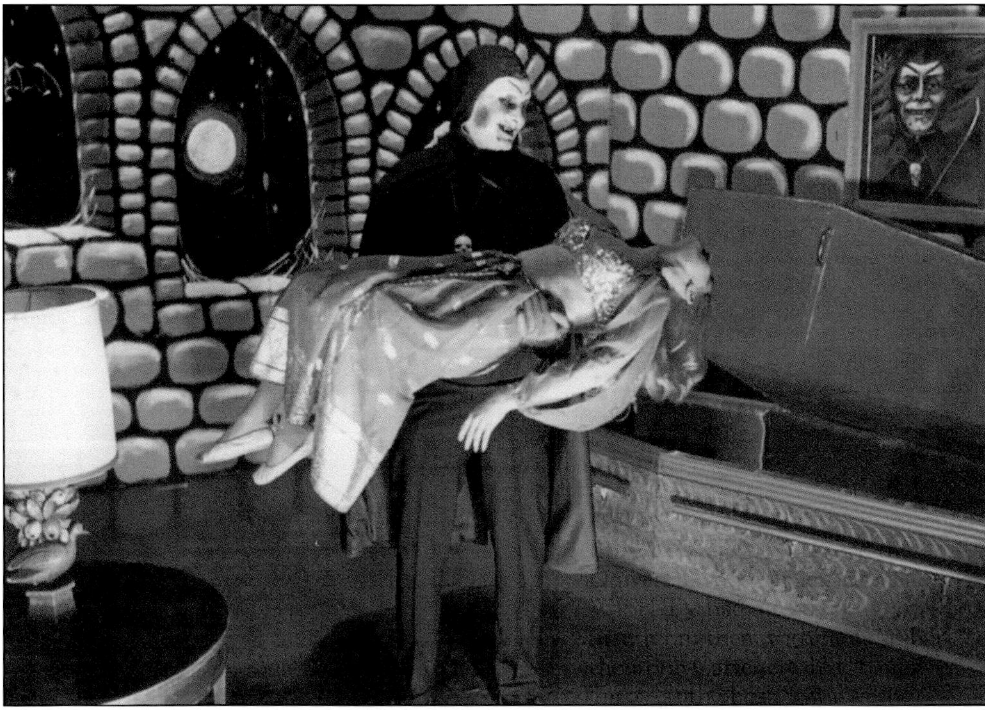

Whereas Selwin was scripted and Sparenberg used a teleprompter (he did not write his own material), Sammy Terry was mostly ad-lib.

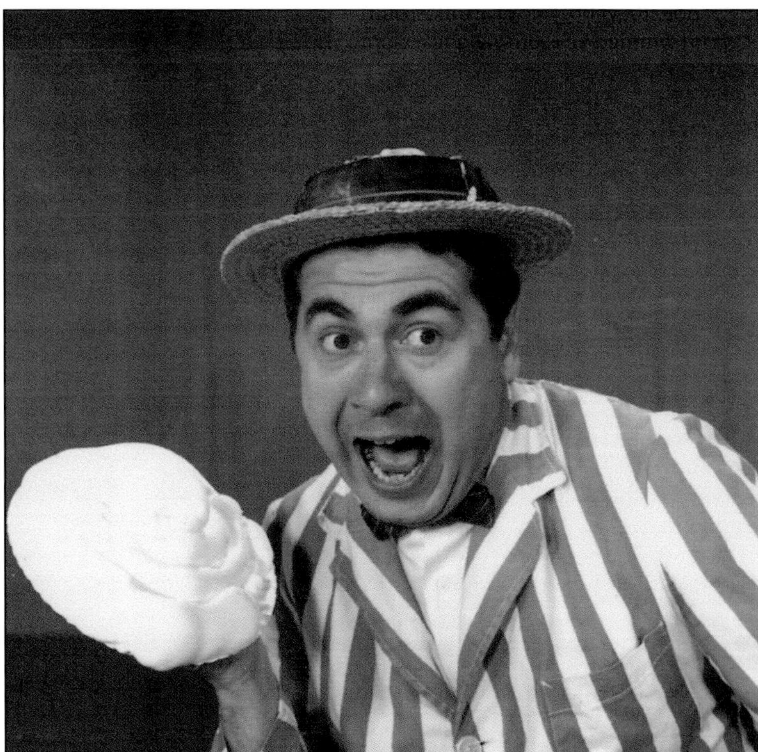

Children's shows were very popular from the onset of television, and local hosts became household names. At WFBM TV, Hal Fryar created a character he named "Harlow Hickenlooper" in 1960. Harlow sang his own peculiar version of "Happy Birthday" for children, always ending with a pie in the face.

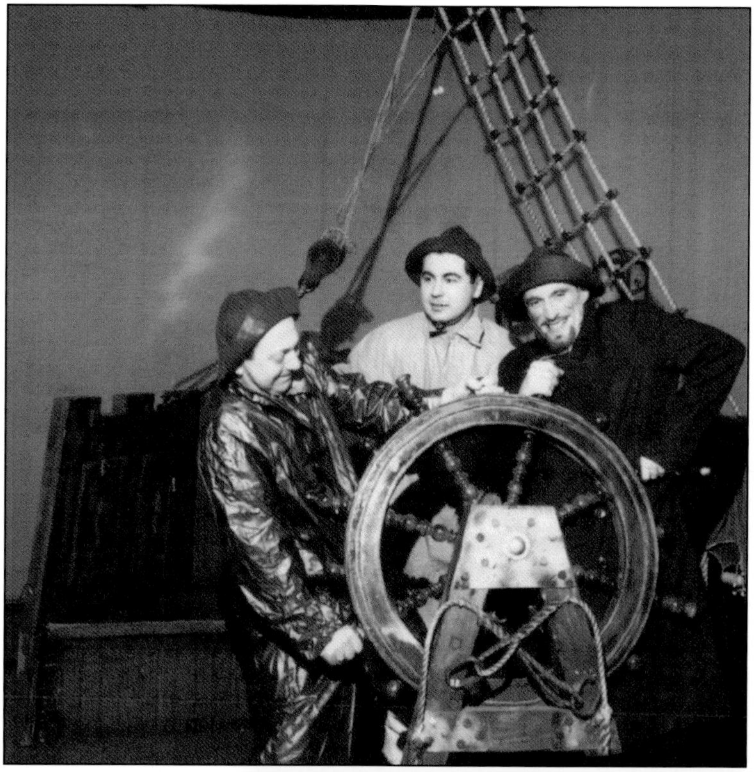

In the beginning, Harlow's cohorts were Jerry Vance as "Cap'n Star" and singing cowboy "Curley Myers." The show was called *All Hands on Deck*. Vance joined WFBM TV as a staff director in 1951. He left in 1968 and moved to Los Angeles, where he became "Seymour," a very popular horror host at KHJ TV and KTLA.

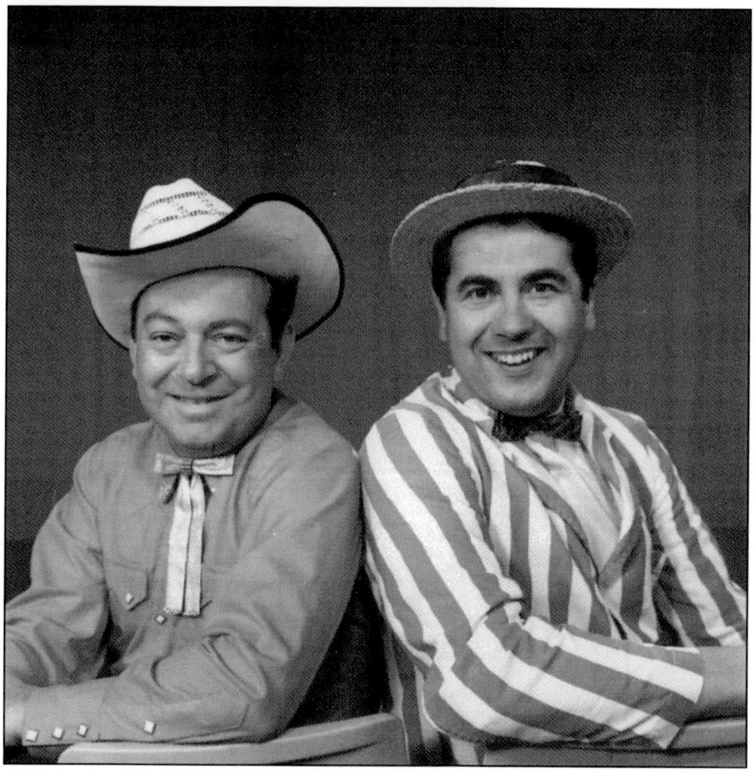

Fryar and Myers continued to host the Three Stooges movies until 1972. After leaving WFBM TV, they continued to do personal appearances for a number of years. Fryar became an account executive and vice president of two advertising agencies and then worked 15 years at Jackson/ Zender Studios in Indianapolis.

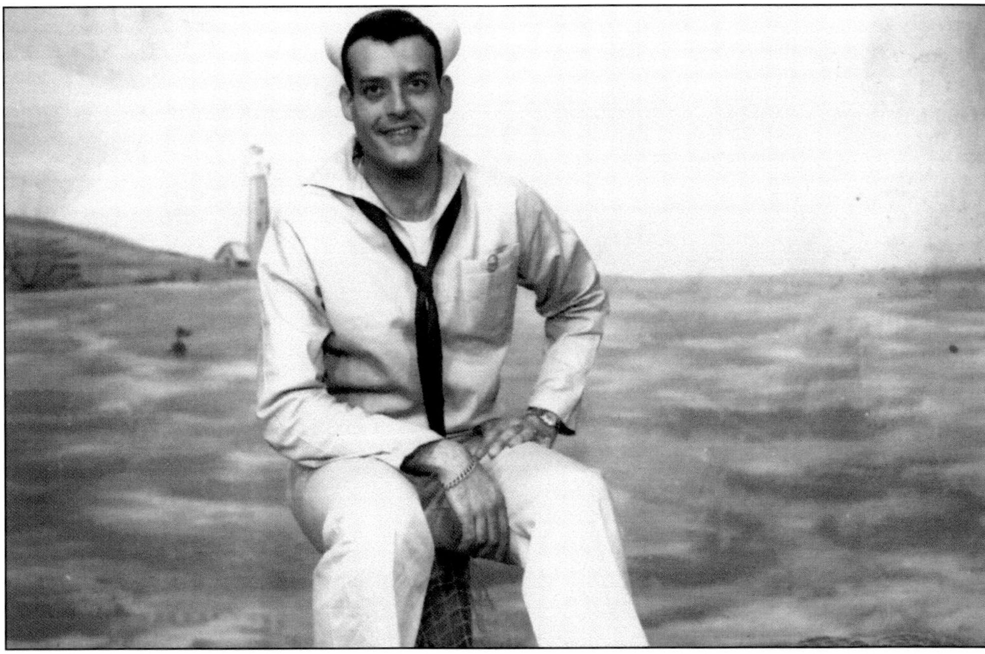

Herb Isaacs was an early children's host at WTTV. He was known on the air as "Happy Herb the Sailor." Isaacs came to WTTV from WTHI TV in Terre Haute in 1959. He was on the air with *Popeye* cartoons six days a week. He left the station in 1965 to become a national sales representative in Chicago.

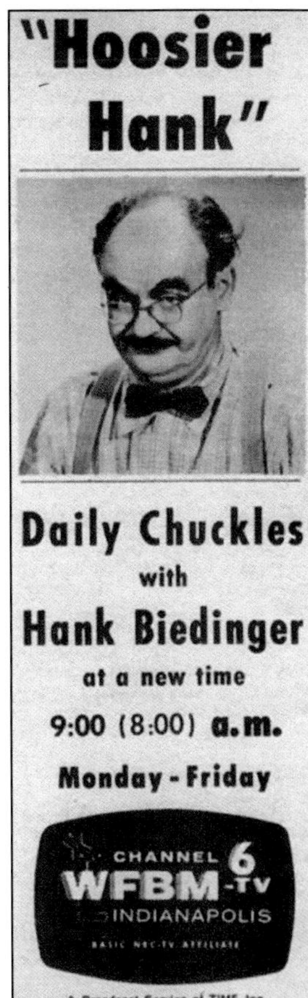

Local amateur actor Henry Biedinger portrayed "Hoosier Hank," a friendly old-time owner of a grocery store. He first appeared on WFBM TV in early 1960. Hank talked to the kiddies before hosting one Three Stooges short. At first, he was seen Monday through Friday for a half hour at 9:00 a.m. Sometime later in 1960, Hoosier Hank was moved to 4:30 p.m. Monday through Friday and thus picked up a much larger younger audience.

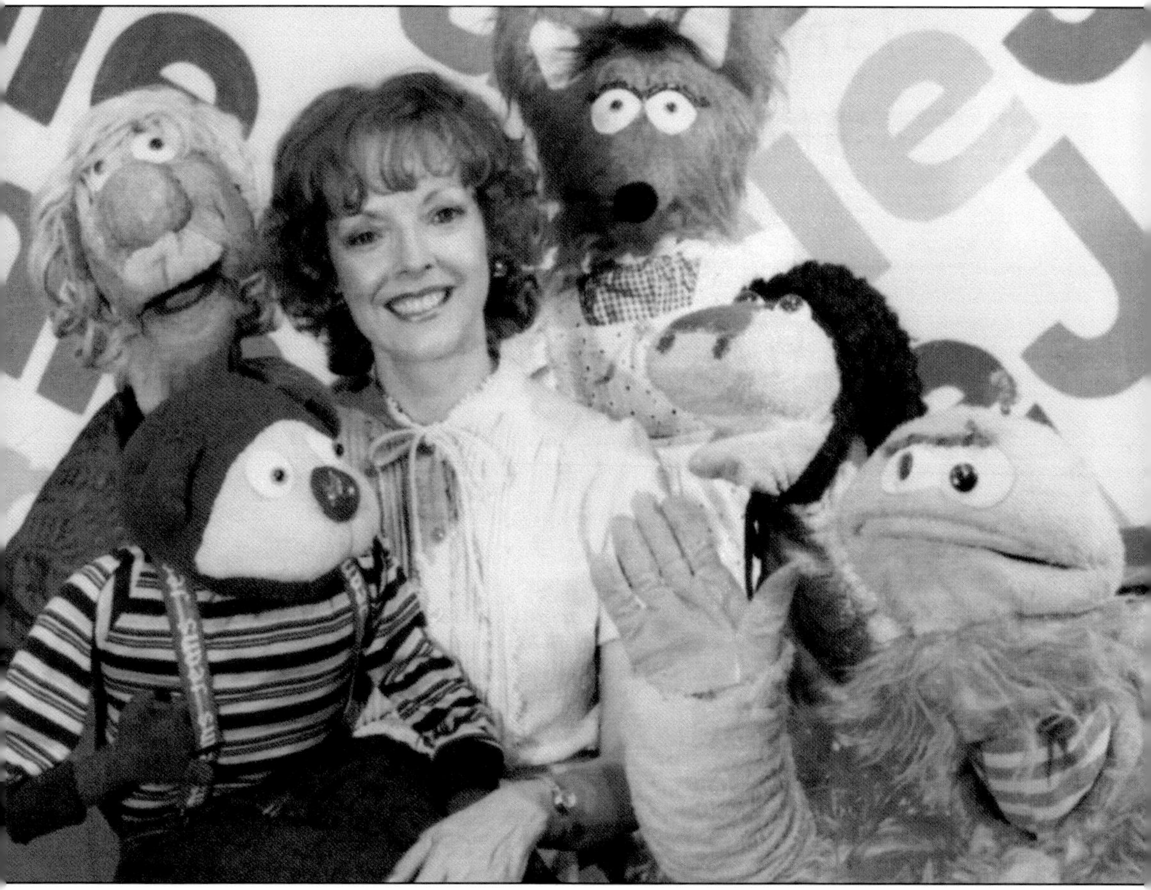

Janie Hodge was a schoolteacher who wanted to have her own TV show. After doing some temporary talent jobs at WISH TV, Janie went to see Don Tillman, the program director at WTTV who was looking for someone to host *Popeye* cartoons.

Janie proved to be the perfect choice, and *Popeye and Janie* was born. Janie would sing, play the ukulele, talk with her puppets, and introduce the *Popeye* cartoon. She had a very successful run from 1963 to 1986.

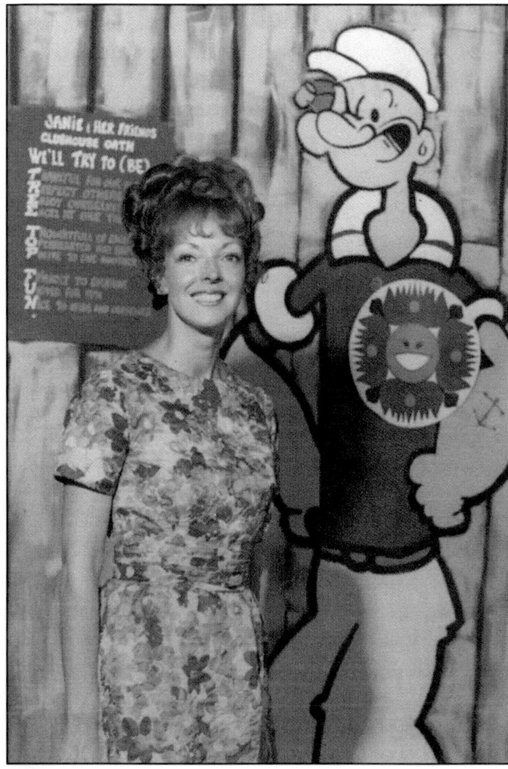

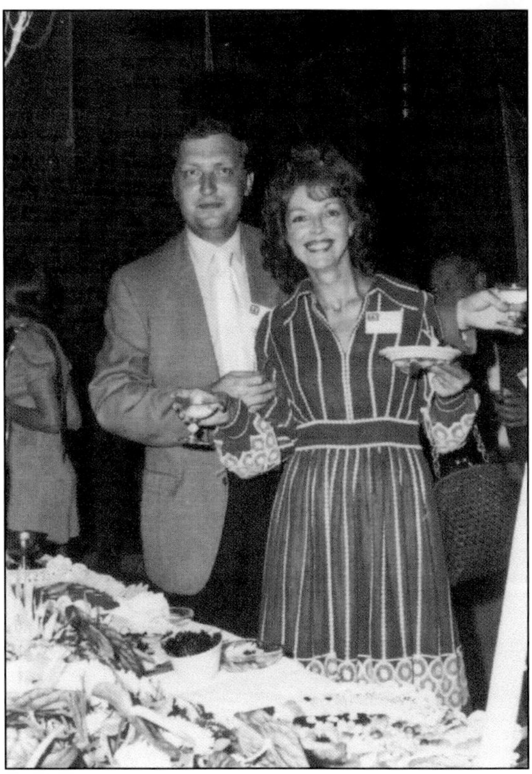

Janie is seen with her WTTV boss, Don Tillman.

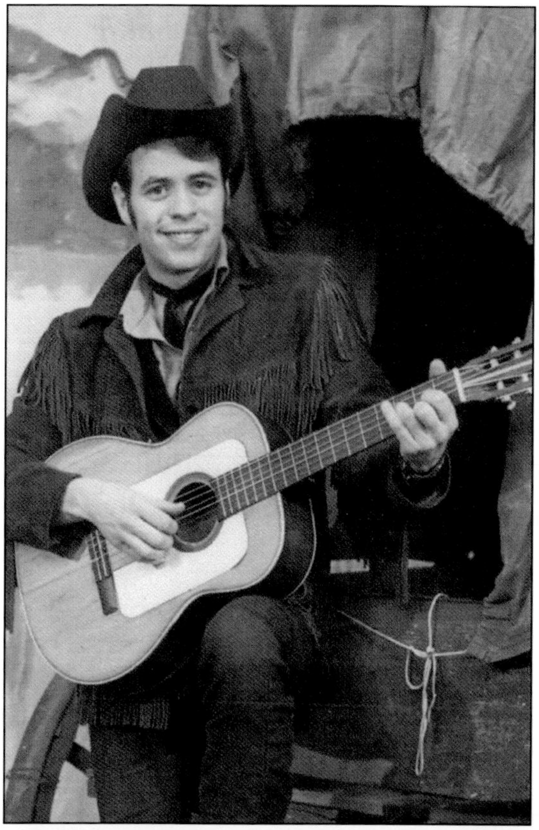

Having been born and raised in Oklahoma, Bob Glaze thought it would only be fair for management to let him be a cowboy. He first appeared as "Cowboy Bob" on Janie Hodge's show, backing up her ukulele with his guitar, and they started singing together. These performances apparently convinced management to let Glaze host his own show. For more than two decades, Cowboy Bob entertained his young audience with "Freckles," "Windjammer," "Skye," and his trusty sidekick, "Tumbleweed."

WISH TV never committed as much airtime to children's shows as other local stations; instead, a local show, *Breakfast with Chuckles*, began in 1954 and was on every Saturday from 9:00 a.m. to 10:00 a.m. It was broadcast with a live audience from WISH TV's 150-seat auditorium studio. Reid "Chuckles" Chapman, who started the popular show on WISH Radio, was a delightful emcee. The show ended in 1956, when Chapman moved to WANE TV in Fort Wayne, where he became general manager.

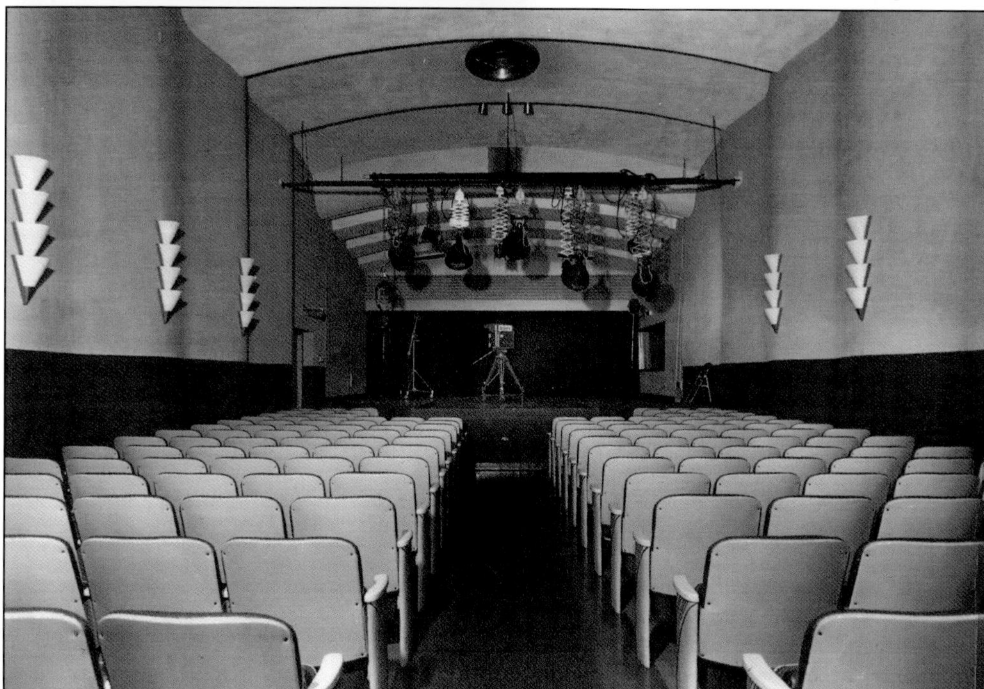

WISH TV tried several children's-show formats in Saturday morning time slots in the 1960s. The first was called *The Kiddie Komedy Hour*, followed by *The Uncle Buster Show*, and then *Clowning Around*. Shown here are Joe Leamon as "Uncle Buster" and Bob Warren as "Ronald McDonald."

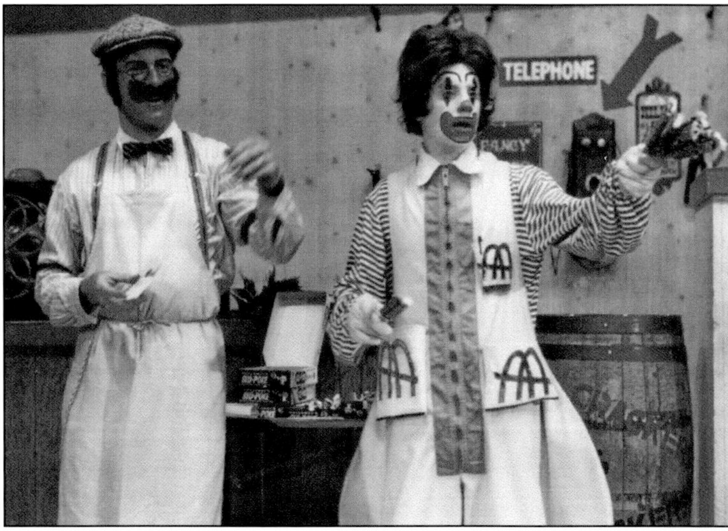

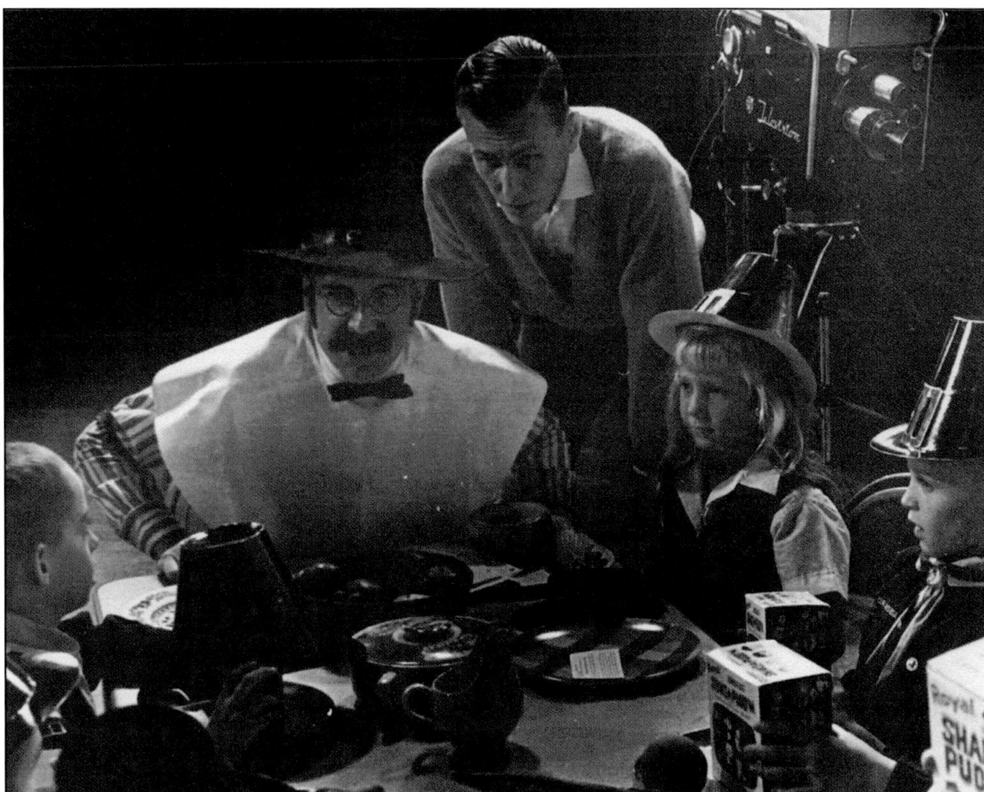

Joe Leamon as "Uncle Buster" teaches children about Thanksgiving; directing Leamon is Dave Smith.

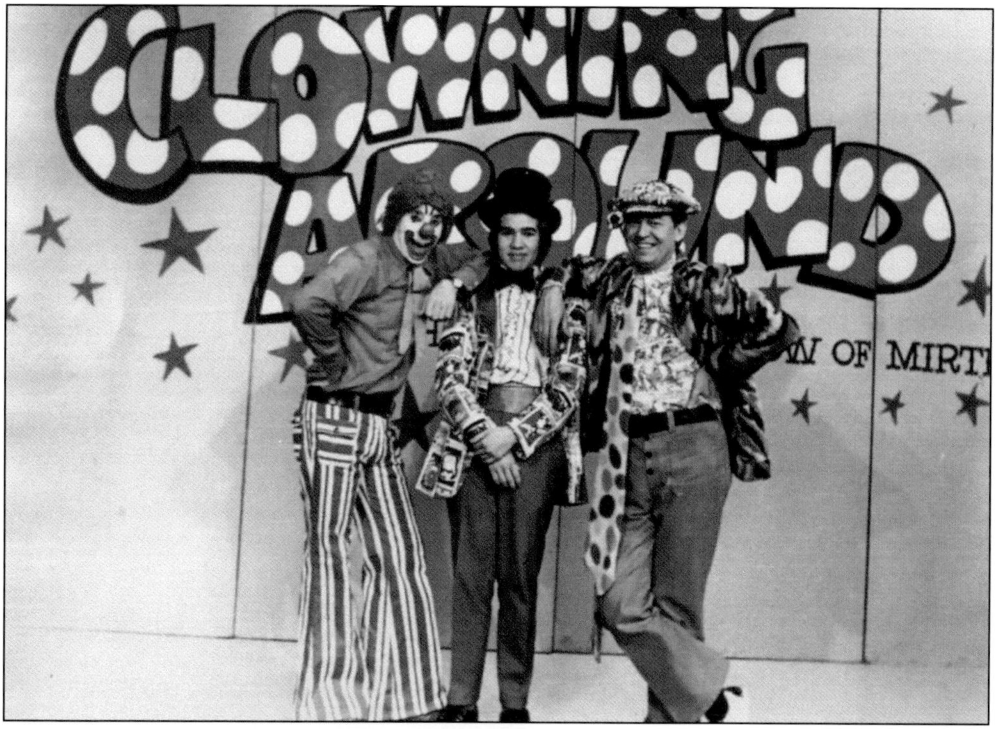

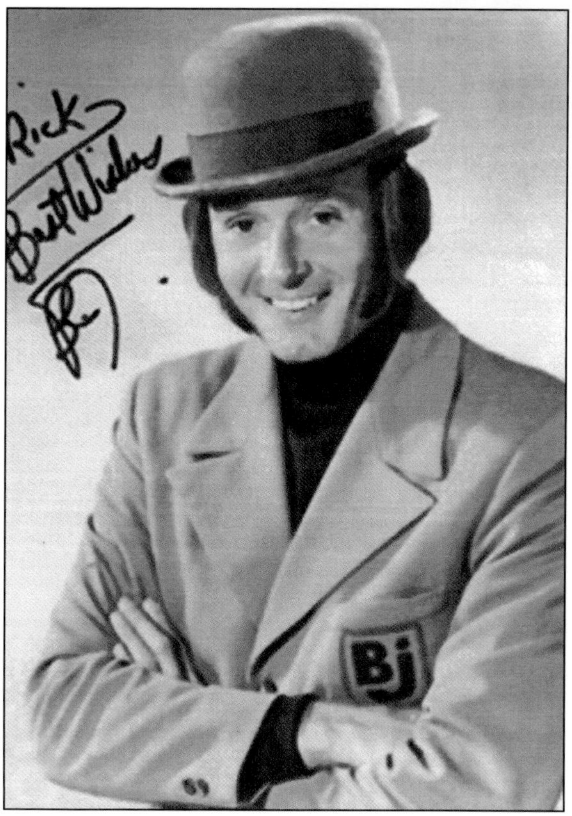

The most successful children's show on WISH TV was *Clowning Around.* It ran from 1968 to 1974. Bob Warren was "Nosey the Clown," Marc Much was the emcee, and Charlie Worpell was "Crazy Charlie," the magician.

Bill Jackson got his start in Chicago as a TV clown named Freckles. He soon moved to Fort Wayne, Indiana, where he was host of a *Popeye and Little Rascals Show.* He came to Indianapolis to work for WLWI TV and began to develop a cast of puppet characters. In 1965, he created *Clown Alley.* His show was so popular that WFLD TV in Chicago hired him. He was on the air with his puppets until 1980, when he moved to California to become a teacher at Cal Arts.

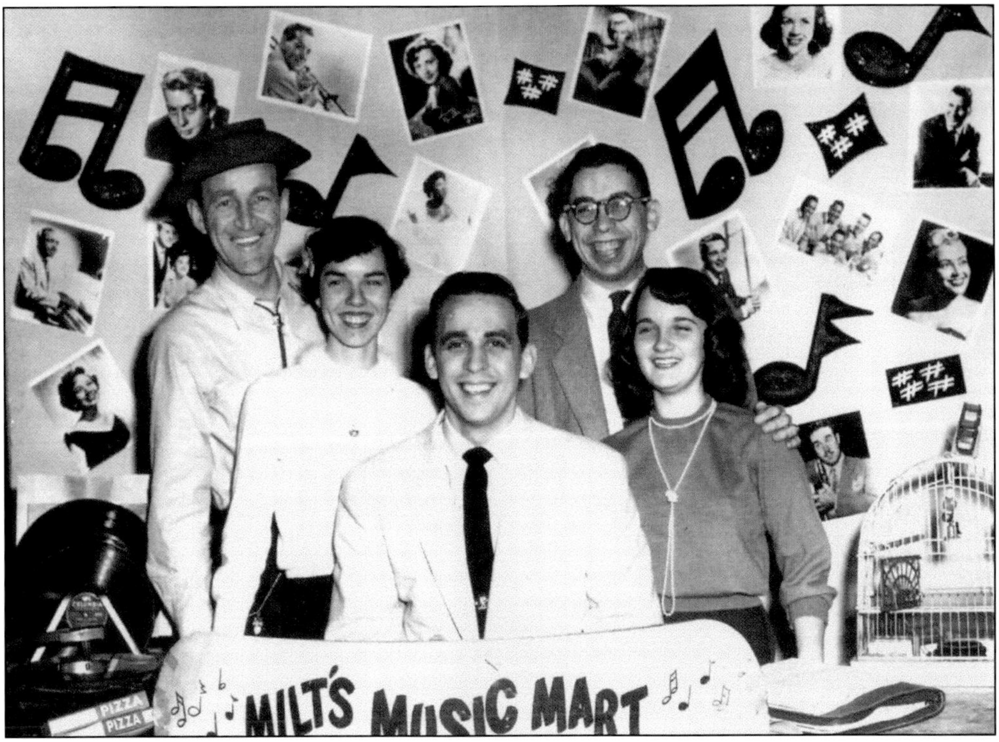

Milt Lewis (center) hosted one of the early *Bandstand*-type programs from the WTTV studios on Hillside Drive in Bloomington. With Lewis are veteran cameraman Earl Sutherlin (left) and Forest Respess (right); the two girls are unidentified.

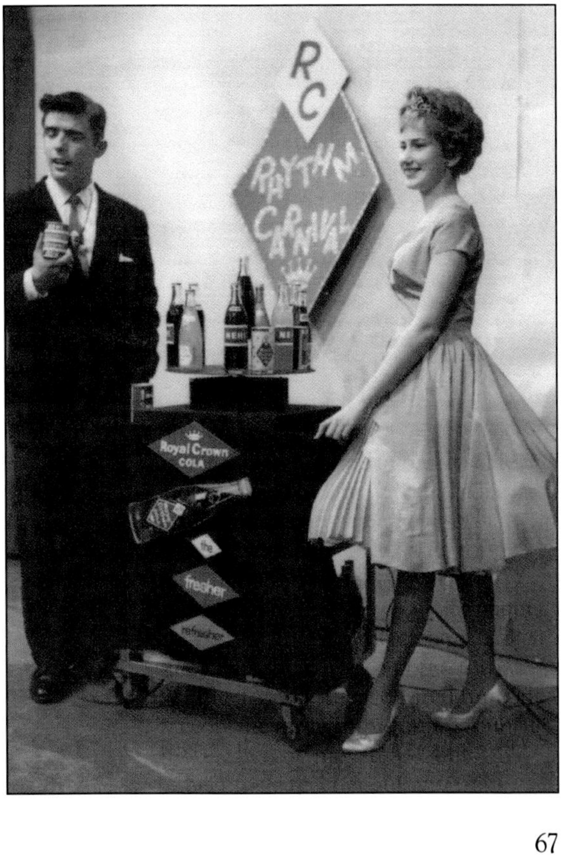

In the 1960s, WISH TV decided to do its own *Bandstand*-type show. WISH Radio had a disc jockey named Dick Summer, and it was decided to put him on television on Saturdays. Since RC Cola was the sponsor, the show was called *Rhythm Carnival*.

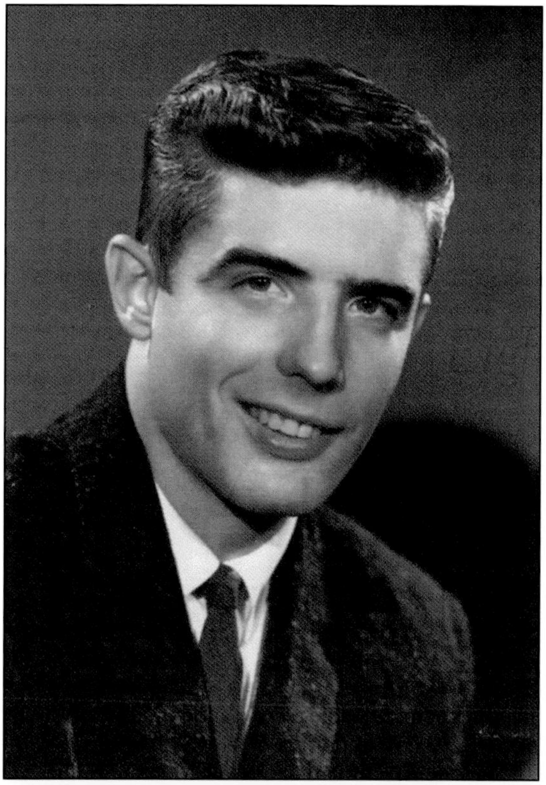

Rhythm Carnival was a hit and ran until Dick Summer was hired away by WBZ Radio in Boston. In the 1960s, Summer was the most popular disc jockey in Boston. Since WBZ was 50,000 watts, he was heard in 38 states and part of Canada. Summer's radio show was unpredictable. He read poetry and short stories, and sometimes he played old radio dramas—listeners loved it.

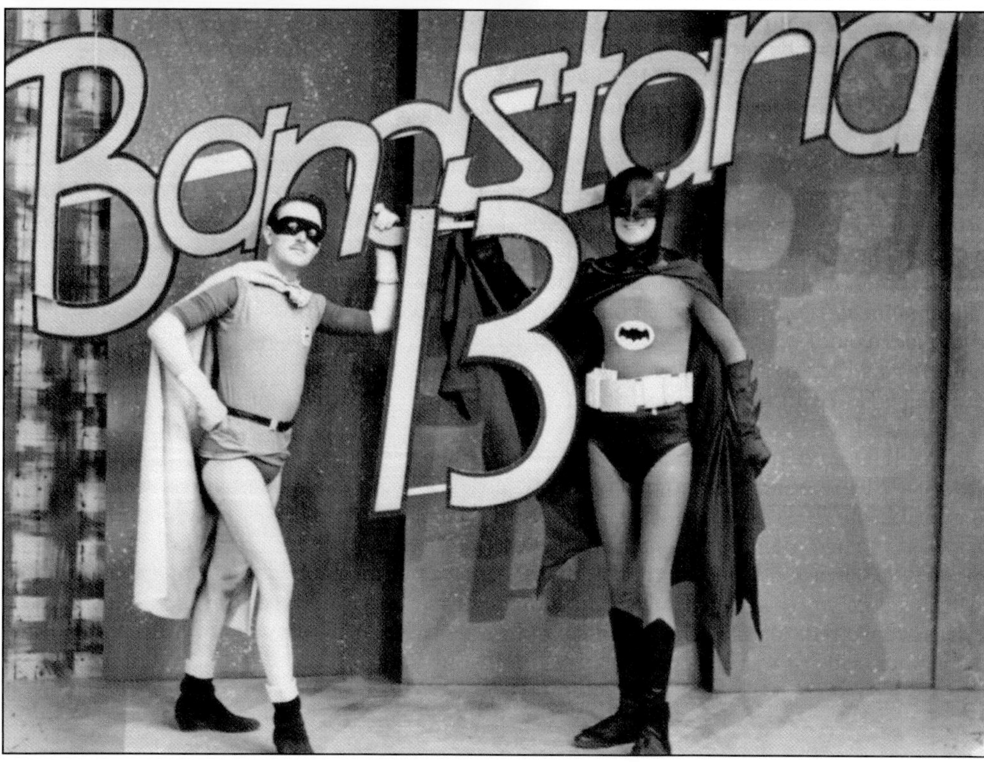

Jimmy Mack had a long career in radio before moving into television. He worked for stations in Chicago, Benton Harbor, and Grand Rapids before settling in Indianapolis, where he worked at WXLW, WIRE, WISH TV, and WLWI TV. He was at WLWI when he created *Bandstand 13.* Here, he is dressed as Batman with Bill McPherson as Robin. The *Batman* show ran on ABC from 1966 to 1968, while *Bandstand 13* ran from 1965 to 1969.

Eight

At Long Last, Educational Television

In 1967, Indianapolis was one of the largest TV markets in the country without the benefit of an educational television station. The FCC had set aside UHF channel 20 for educational use in the Indianapolis market. In 1967, there were 156,000 Indianapolis homes that had TV sets capable of receiving a UHF signal.

Dr. I. Lynd Esch, president of Indiana Central College, headed a group that had incorporated as the Metropolitan Indianapolis Television Association, Inc. (MITA). It had applied for channel 20 in 1965, but the FCC had not granted it. This was largely due to a dispute involving WTTV, WISH TV, and WLWI TV. WTTV wanted to move closer to Indianapolis and had asked for a tower site on West Ninety-sixth Street. As an inducement for the FCC to approve the move, WTTV offered $350,000 in cash and equipment to help channel 20 get on the air.

WISH TV and WLWI TV offered to match the WTTV bid if the FCC did not approve WTTV's plan for a tower move. The FCC rejected WTTV's plan, and the station had no choice but to have its tower and transmitter stay in Trafalgar, Indiana. WISH TV also offered channel 20 the use of its new tower that the company hoped would be built northwest of Indianapolis, where channel 6 and 13 had their towers. WISH TV built its tower next to its radio tower on the east side of Indianapolis, and this meant it could not deliver a good signal in the Lafayette area. This opened the way for a UHF station in Lafayette that became a CBS affiliate. The Lafayette station and Sarkes Tarzian strongly objected to the WISH TV tower move.

John Burkhart, president of College Life Insurance Co., filed an application for UHF channel 59 and stated it would operate as an educational station. He proposed to use WTTV's tower if it was moved to Ninety-sixth Street and Ditch Road. WISH TV and WLWI TV quickly offered tower space as well. Burkhart did not get channel 59 on the air, as it became apparent that WFYI TV would occupy channel 20. Burkhart's wife, Ardath, led an all-out, block-by-block, door-to-door assault to raise $350,000 to get the station on the air. WFYI TV signed on October 4, 1970, with a staff of nine employees, three black-and-white cameras, and a signal that barely reached downtown Indianapolis. It was the city's first UHF TV station and the first to offer commercial-free programming.

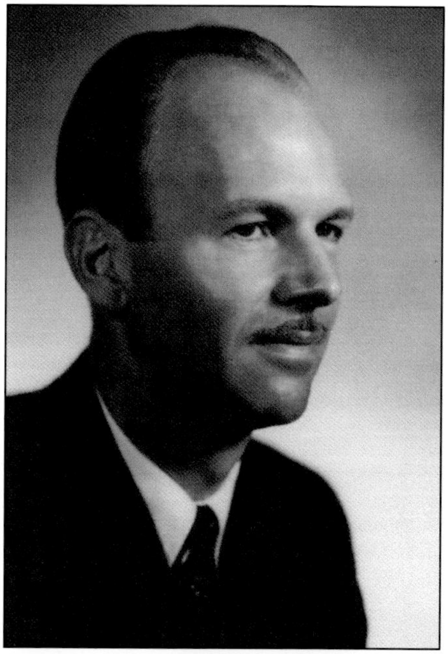

In 1965, Warren Wright, former program manager at WFBM TV, was given the task of heading an educational television station with no cameras, no studios, and no money—just call letters WFYI TV and a worthy cause. The station finally went on the air on October 4, 1970. It was initially housed on the grounds of the Indianapolis Museum of Art. When WTHR TV moved into a new building at 1000 North Meridian Street, WFYI TV occupied its old building. Another move occurred when larger quarters in the Indiana Gas Building at 1630 North Meridian Street became available. Over the years, WFYI TV has garnered an impressive number of regional Emmys for its locally originated programming.

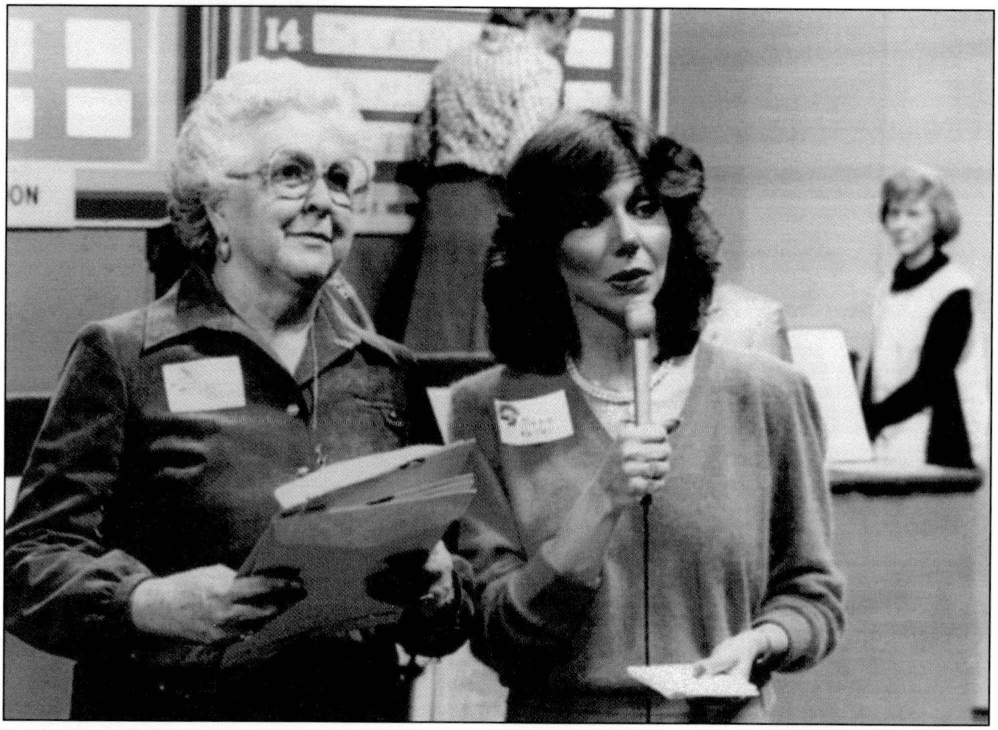

In the late 1960s, Indianapolis was the largest city in the nation where programs such as *Mister Rogers* and *Masterpiece Theatre* were unavailable for viewing. The only way to watch *Sesame Street* was on a commercial TV station. In December 1969, Ardath Burkhart (left) became interested in WFYI TV when she found that the male backers of the channel were unable to raise $350,000 to get the project started. Ardath Burkhart recruited thousands of women who became known as "Ardath's Army" and raised the money in less than one year.

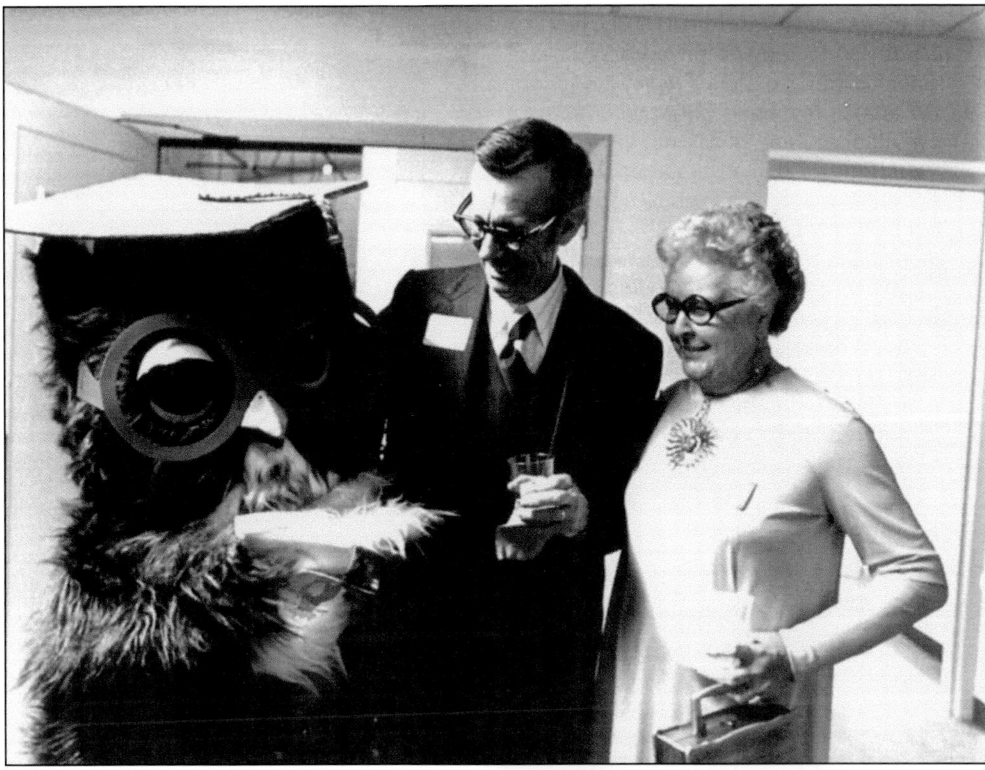

When Warren Wright left WFYI TV in 1972, Frank Meek, the former commandant of the Defense Information School at Fort Benjamin Harrison in Indianapolis, became vice president and general manager. Meek retired in 1989 after guiding the station successfully during his tenure, which included the addition of NPR radio station WFYI FM.

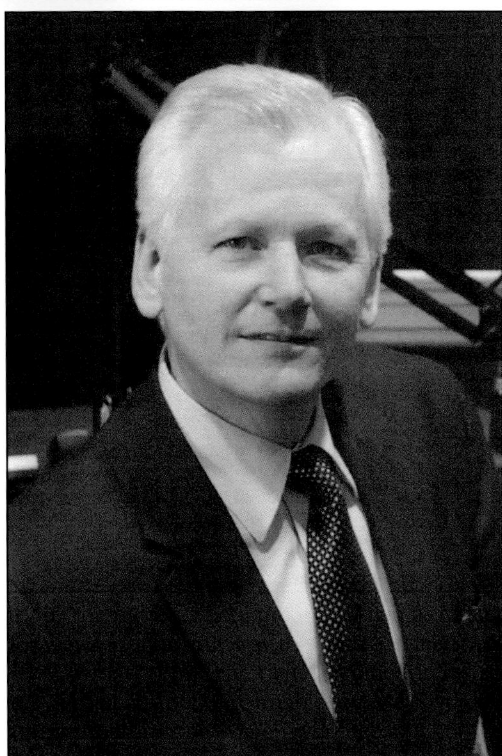

In 1988, Lloyd Wright joined WFYI TV as director of program production. The following year, he was appointed WFYI's president and chief executive officer. Wright had more than three decades of experience in public-service media. Prior to his arrival at WFYI, Wright served for six years as broadcast operations manager of WTTW (Chicago), was producer/director/writer for the Indiana University News Bureau, and director of instructional broadcasting for the Indiana Department of Education. In 2008, under his leadership, he relocated WFYI Public Media to a state-of-the-art facility in the former Indiana Gas Building on "Broadcast Row."

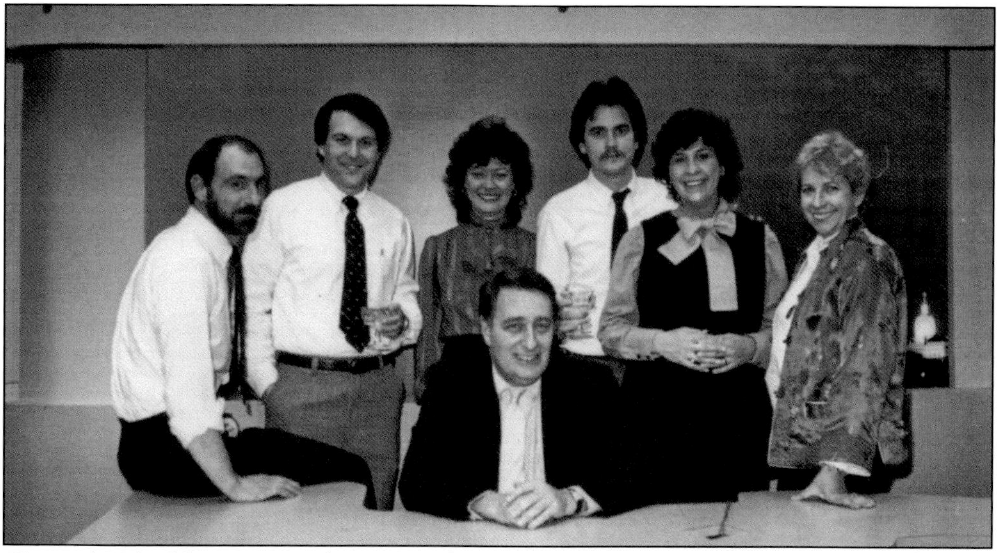

Channel 20 had been assigned to Indianapolis for quite some time before it went on the air. Similarly, channel 59, a commercial UHF frequency, was dormant for a number of years. This all changed when former channel 13 general manager Chris Duffy forged an agreement with the Simon brothers and Ron Palamara, president of Anacomp. WPDS TV went on the air on October 1, 1984. From left to right are Jim Borgioli, John Newcomb, Carole McPherson, Mike Sanders, Kim Sanders, Beth Sullivan, and Chris Duffy (seated).

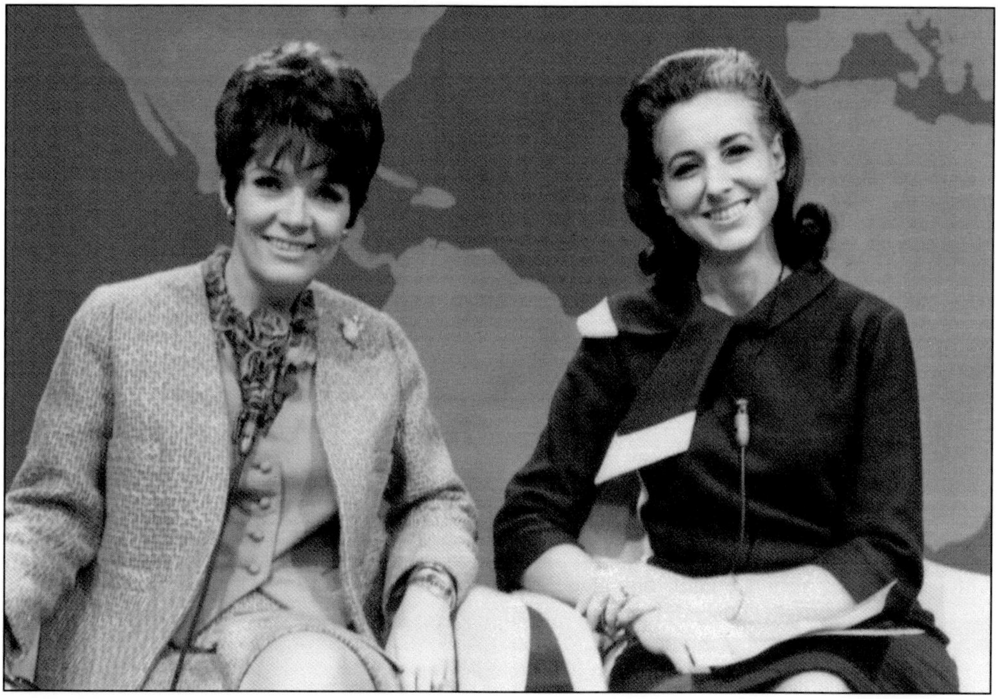

In the 1960s, women began to appear in newsrooms. Faith Levitt was the women's director at WISH TV from 1963 to 1971. As such, she appeared daily on the noon news. She pioneered the coverage of subjects that were considered controversial at that time: unwed mothers, abortion, the La Leche League were among them. She is shown here with actress Polly Bergen (left).

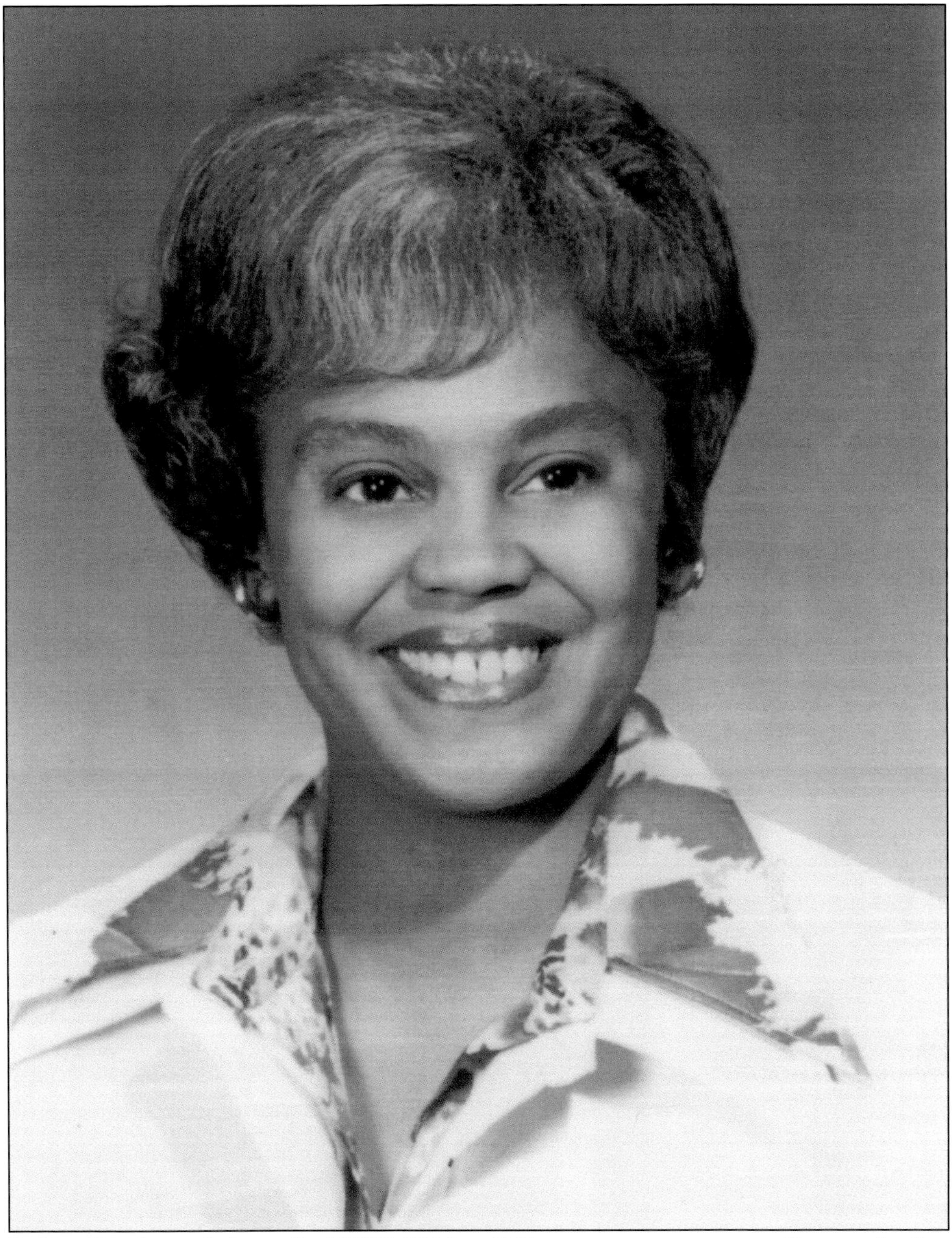

Barbara Boyd joined the staff of WFBM TV in 1969 as a consumer reporter. She was the first African American female to anchor a news program. Boyd's early broadcasts on social and economic issues quickly earned her a loyal following. She received numerous awards throughout her career. The reporting of her own mastectomy from a hospital bed resulted in her being named "Woman of the Year" by the American Cancer Society. She was inducted into the Black Journalism Hall of Fame, the Indiana Journalism Hall of Fame, and the Indiana Broadcast Pioneers Hall of Fame.

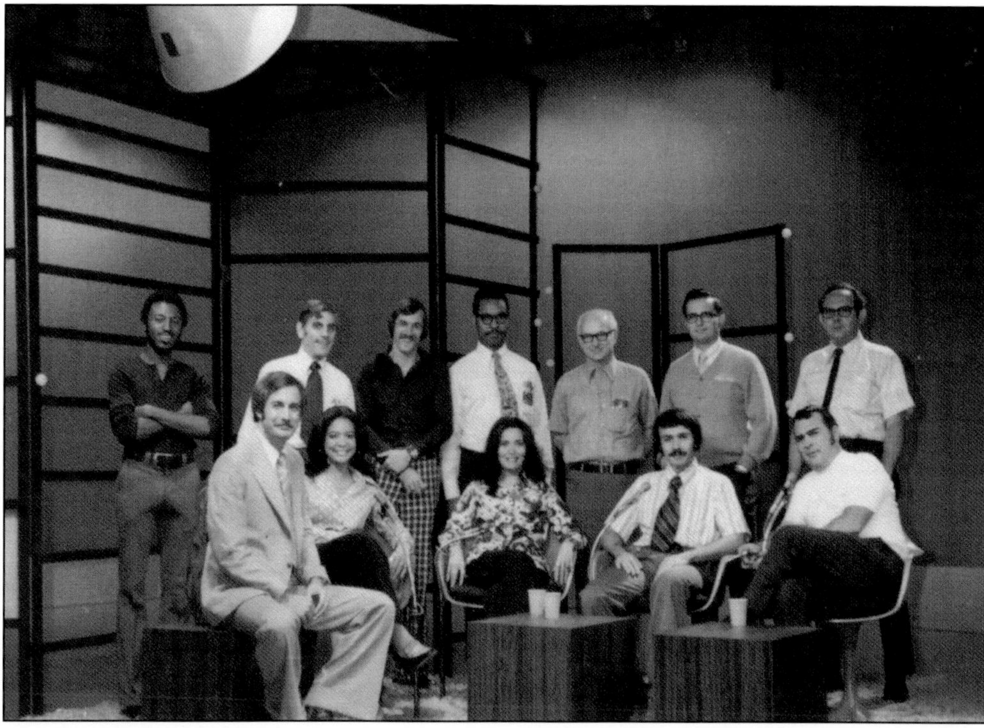

Women began to host local talk/interview shows. One of the first in Indianapolis was *Indy Today* on WISH TV and featured Janet Langhart as the host. Until 1993, no local TV station had a regularly assigned African American anchor on any weeknight newscast. However, in 1972, *Indy Today* became the first local daily program of this type to be hosted by an African American woman. Seated second to the left is Langhart, and to the right of her is producer Marti McPherson. Langhart left in 1973 when she was hired away by WCVB TV in Boston. Alpha Blackburn, a prominent local African American personality, took over the show.

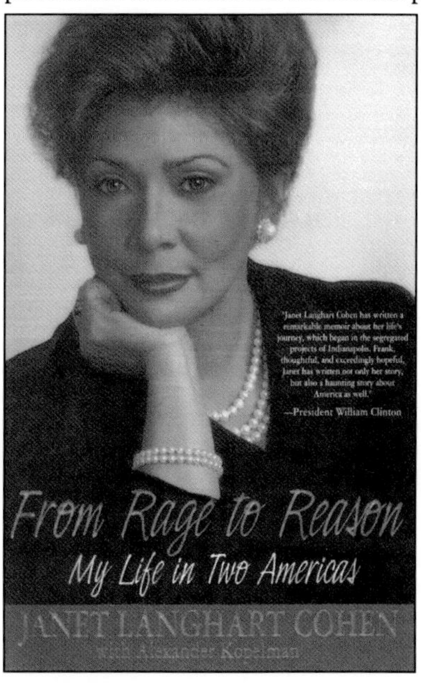

Langhart became cohost with senior anchor John Willis on WCVB TV with a new show dubbed *Good Morning*. It was an immediate success and became syndicated in 72 markets. Langhart was the first African American woman in the United States to do a nationally syndicated talk show. After five years in Boston, Langhart moved to the NBC program *America Alive!* The show did not last, and she joined A.M. *New York* in 1979. After a few more stops, she went back to Boston and rejoined WCVB TV. In 1996, she married William Cohen, the former senator and secretary of defense in the Clinton administration.

Carolyn Churchman was a longtime Indianapolis radio and television personality. Her career spanned more than 55 years, starting in radio in Columbus, Ohio, in the 1940s while attending Ohio State University. She moved back to her Indianapolis hometown where she worked at almost every station in the market. She had her own TV show on WFBM TV, WTTV, and W A V TV. She hosted such popular shows as the *L.S. Ayres TV Shopping Guide* and WFBM Radio's *Coffee at the Marott*. For eight years, she broadcast live from the legendary L.S. Ayres Tea Room.

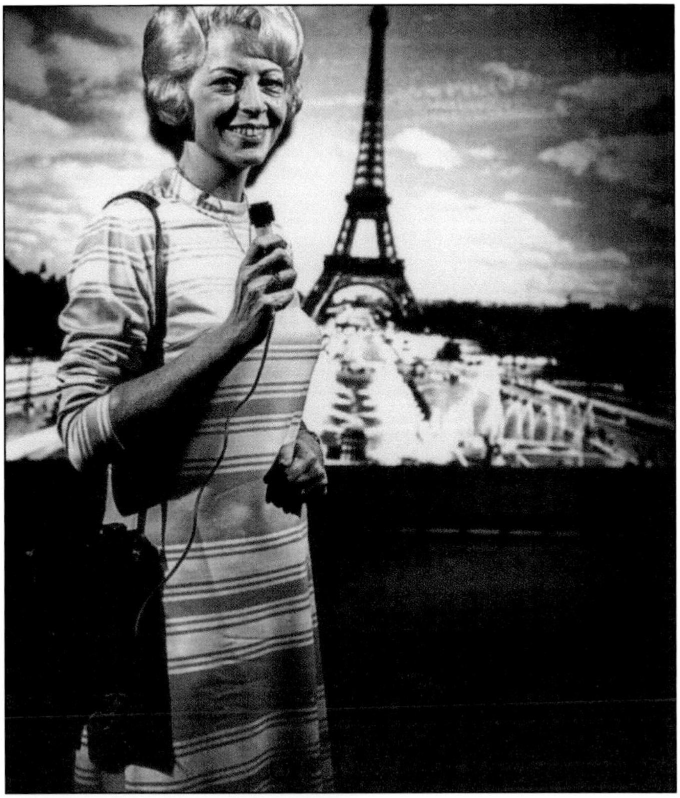

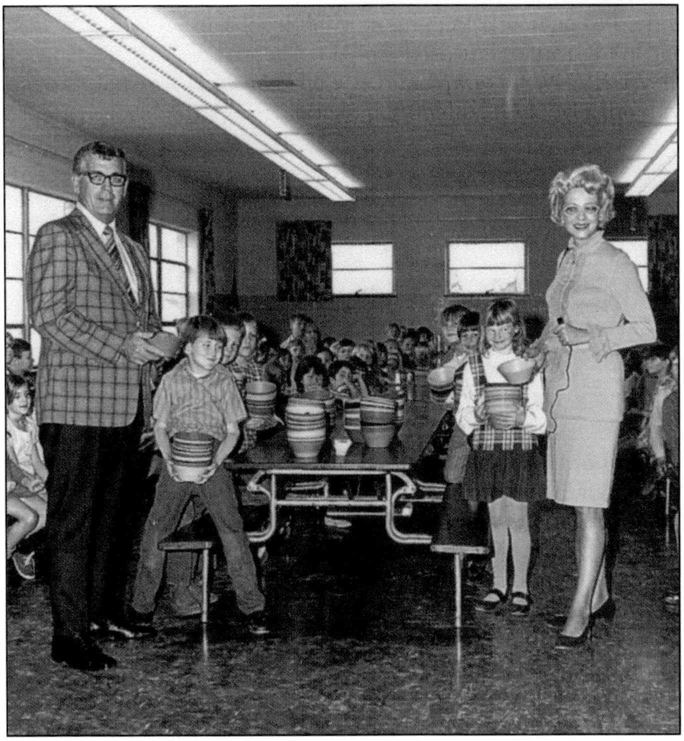

Carolyn Churchman frequently did her shows on location. Here, Joe Piggott, a fellow WFBM TV employee, joins Churchman in this remote telecast where children are showing off their pottery. Churchman was awarded the Francis Wright Award, recognizing her pioneering work in the field of communications from the Indianapolis Professional Chapter of Women in Communications. In 2001, she was posthumously inducted into the Association for Women in Communications Hall of Fame.

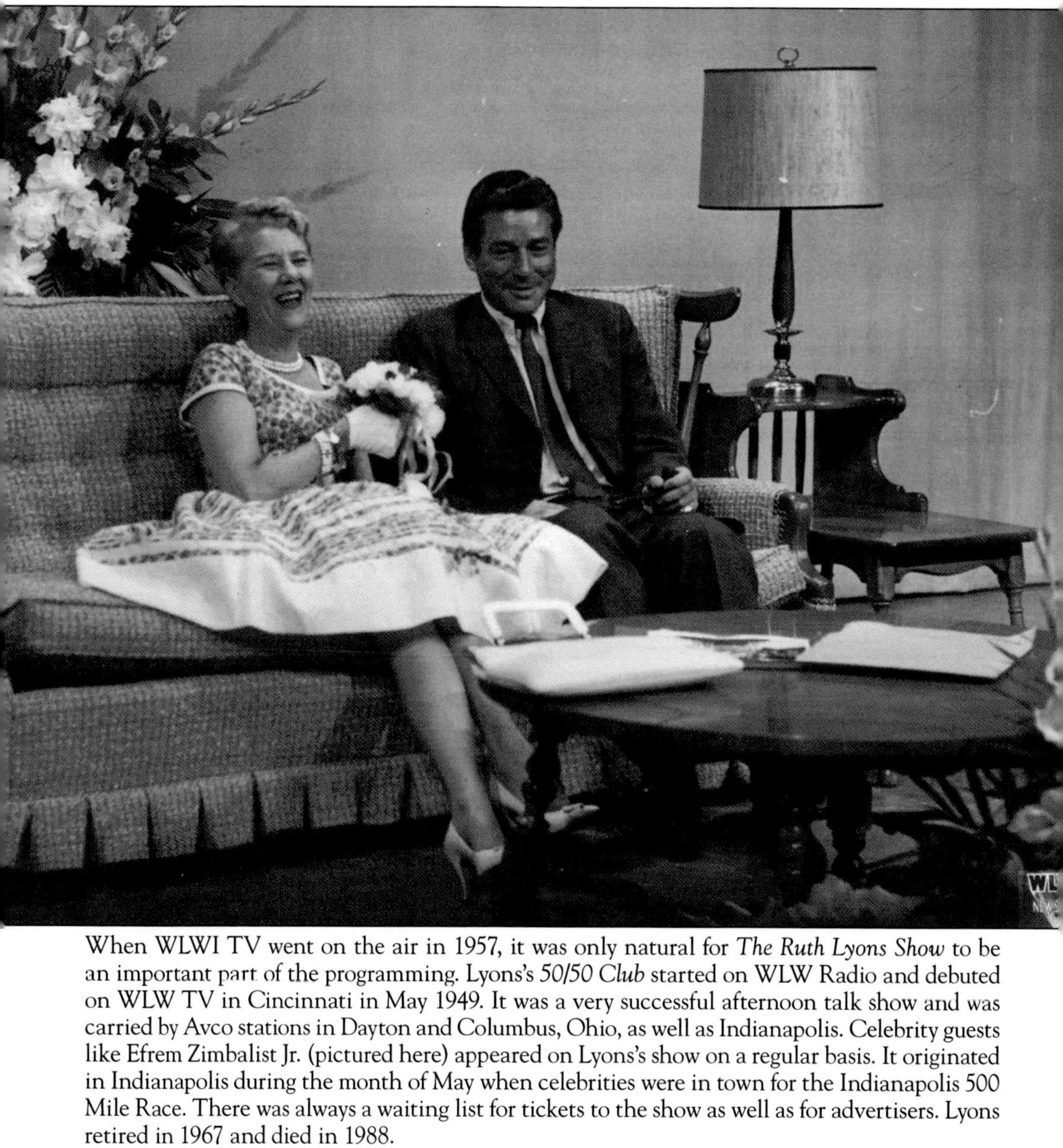

When WLWI TV went on the air in 1957, it was only natural for *The Ruth Lyons Show* to be an important part of the programming. Lyons's *50/50 Club* started on WLW Radio and debuted on WLW TV in Cincinnati in May 1949. It was a very successful afternoon talk show and was carried by Avco stations in Dayton and Columbus, Ohio, as well as Indianapolis. Celebrity guests like Efrem Zimbalist Jr. (pictured here) appeared on Lyons's show on a regular basis. It originated in Indianapolis during the month of May when celebrities were in town for the Indianapolis 500 Mile Race. There was always a waiting list for tickets to the show as well as for advertisers. Lyons retired in 1967 and died in 1988.

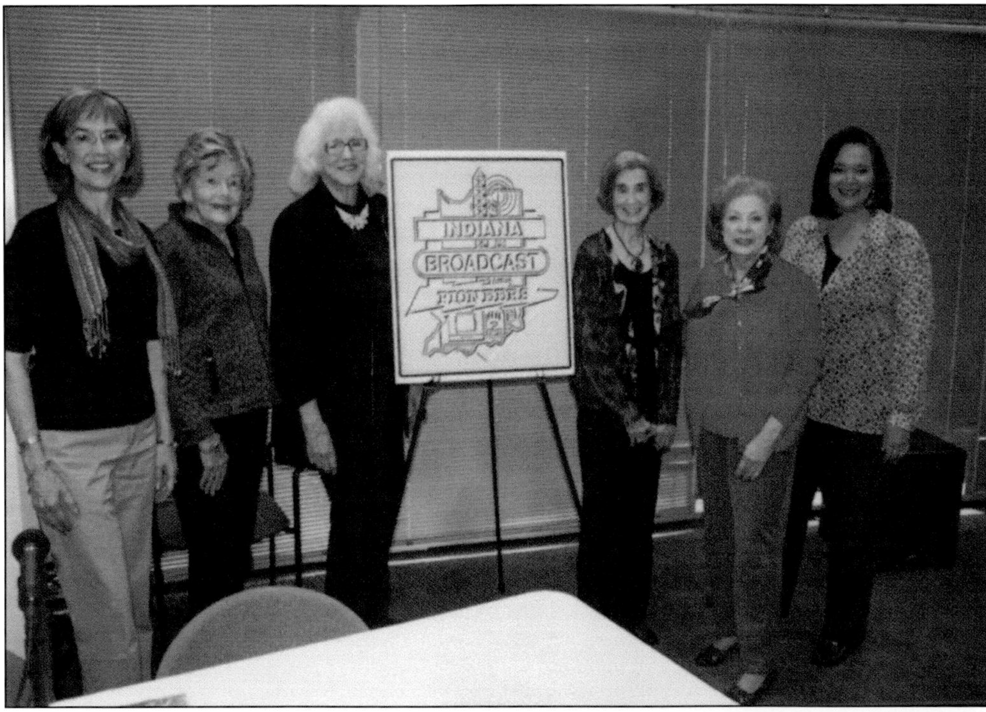

The Indiana Broadcast Pioneers scheduled a luncheon in September 2011. The featured attraction was a panel of "pioneer" broadcast women. From left to right are Kit Field Kruger, Kay Field, Faith Levitt, Ann Wagner Harper, Linda Lupear, and Angela Cain of WTHR TV, who was the moderator.

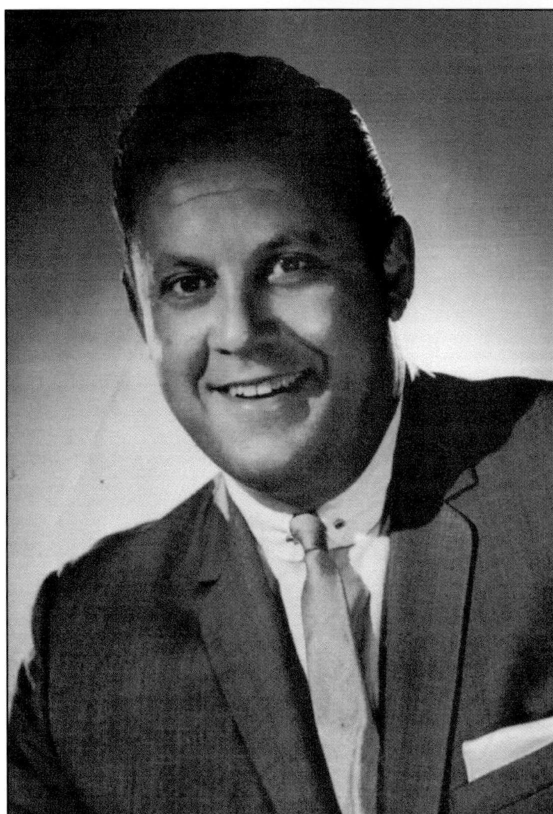

In 1965, WLWI TV decided to start a daily talk/variety show featuring a comic named Don Melvoin. Melvoin began his career hosting a children's show in Traverse City, Michigan, in the 1950s. He then began hosting *The Don Melvoin Show.* The latter show prompted the station to hire Melvoin to host a show by the same name. It ran daily on WLWI TV from 4:00 p.m. to 5:00 p.m.

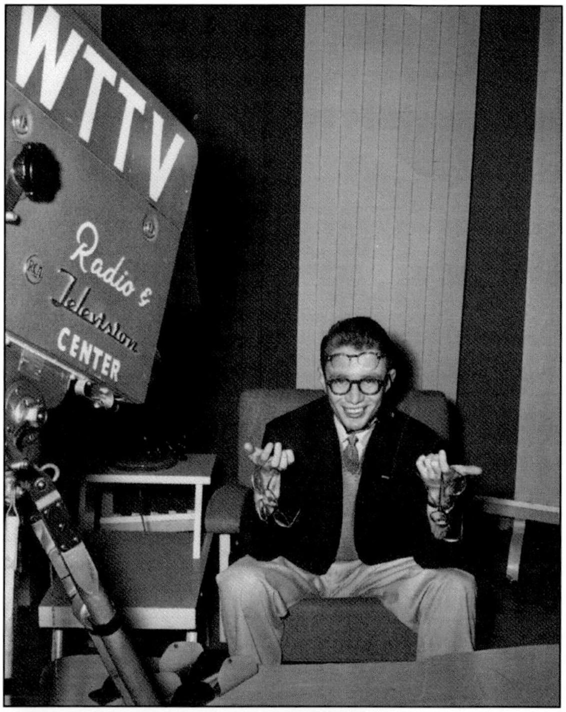

Before he was a weatherman, Stan Wood was a very versatile announcer/host on WTTV. One of his jobs was to host an afternoon movie entitled *Feature at Five*. One day, Stan received a package from a faithful viewer who happened to be an optometrist, and he sent Stan a half-dozen pairs of glasses.

When Stan moved to WISH TV, he became an audience favorite as a weatherperson. His unique weather van traveled across the channel 8 coverage area, bringing on-the-spot weather forecasts from a variety of locations.

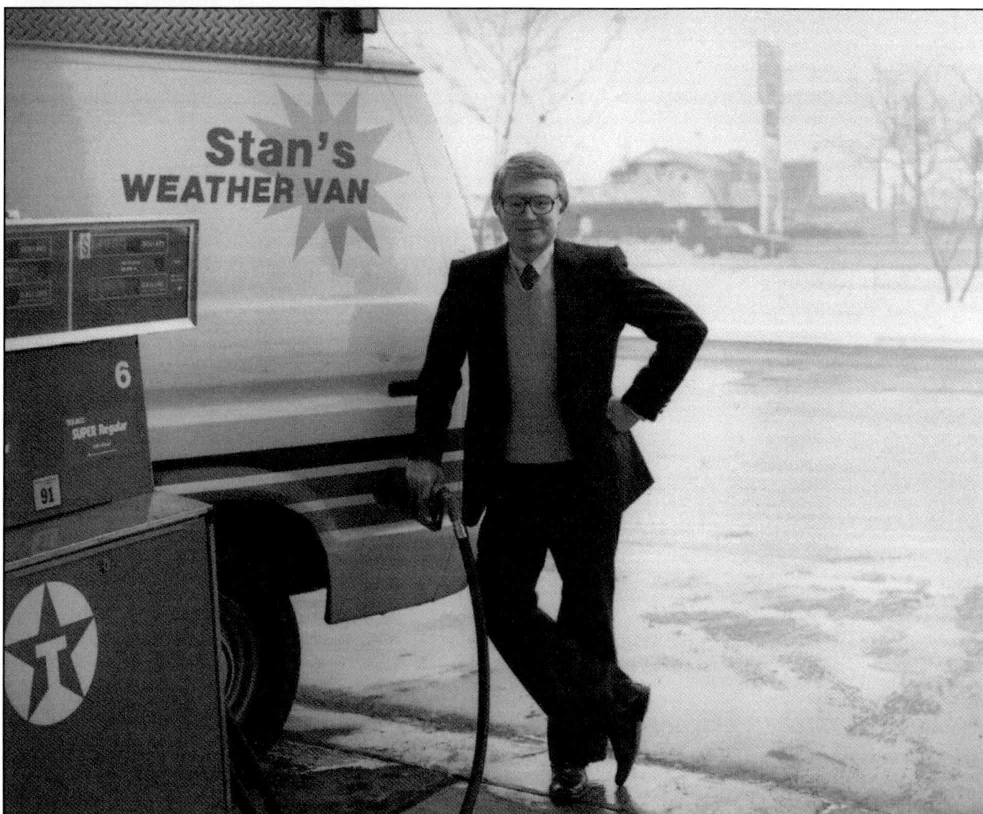

Bob "Swoop" McClain was a popular weatherman for WRTV. He was noted for his wide hand and arm "swoops" across the weather map to show how the fronts were moving.

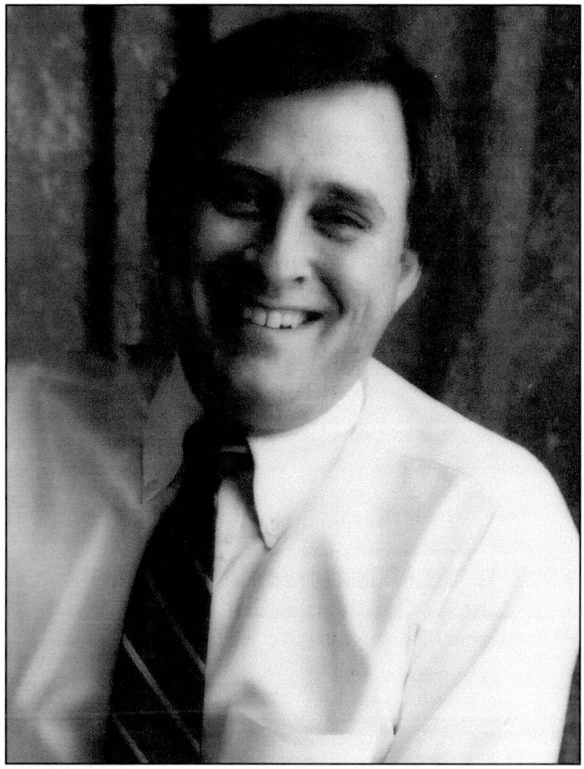

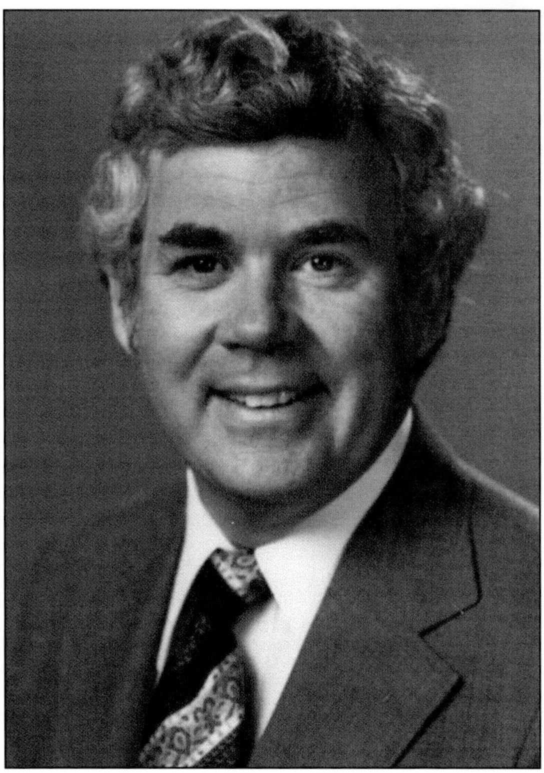

Bob Gregory joined WTHR TV in 1972. In addition to being an excellent meteorologist, he is known for his on-air crack-ups and his "Coats for Kids" campaign. Gregory's campaign has provided 170,000 coats for children in need since 1986. He semiretired in 2000.

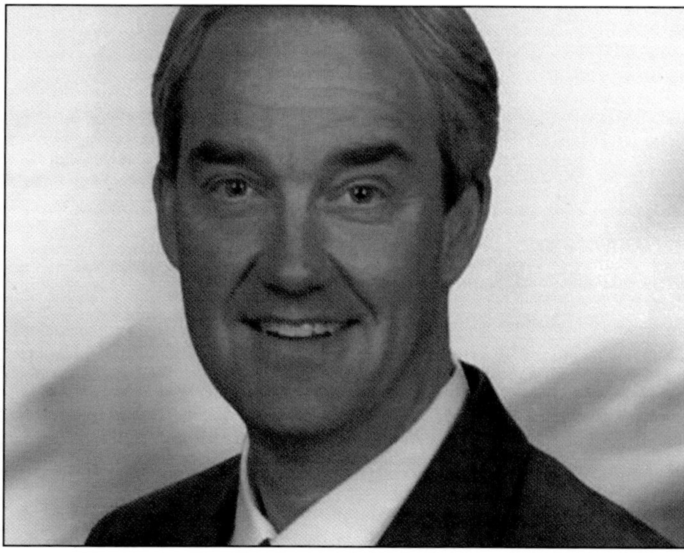

Kevin Gregory got his easy-going weather-reporting style from observing his father. Kevin joined WRTV in 1989 after his father was well established in the market. The father-son combination competed with each other for over 10 years. Kevin did weather for WCIA TV in Champaign, Illinois, and WLFI TV in Lafayette, Indiana, before coming to WRTV.

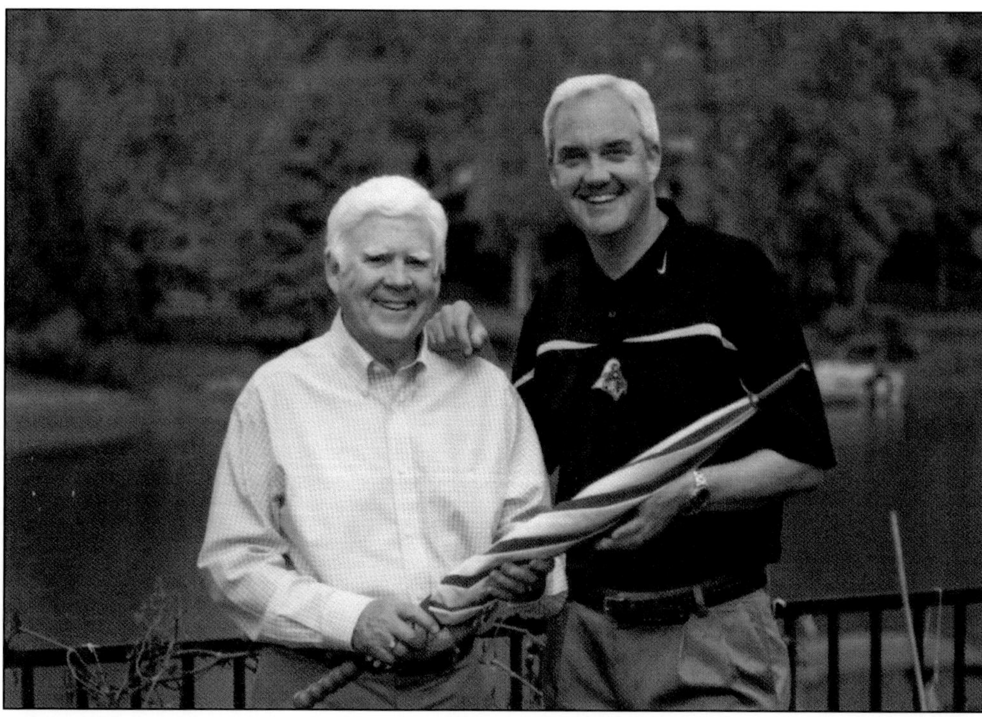

Bob and Kevin Gregory may be the only father-son weathermen competing in the same market in the history of television.

Nine

THE NETWORK SWITCH

In 1975, Avco, owners of WLWI TV, received an offer to sell the station from WBNS, Inc., in Columbus, Ohio. WBNS was part of a family-owned corporation that included a radio and TV station in Columbus as well as the *Columbus Dispatch* newspaper. On March 2, 1975, Avco accepted an offer of $17.65 million, and the call letters were changed to WTHR TV. The Wolfe family, with only two television stations, became the closest thing to local ownership of a TV station in the Indianapolis market.

It is not often that television stations switch network affiliations; however, this did occur in the Indianapolis market in 1979. At this time, the ABC network was on a roll. It once was the weakest of the big three networks, and as such, its affiliations were with weaker stations. It decided to use this opportunity to upgrade its affiliations. It embarked on a national campaign to drop its weakest affiliates by offering an ABC affiliation to the NBC-affiliated stations, since NBC was now the weakest network.

At this time, WTHR TV (formerly WLWI TV) was the weakest network-affiliated station in Indianapolis. It was affiliated with ABC, the strongest network. WRTV (channel 6) was an NBC affiliate with strong local ratings due largely to its local news dominance. WRTV dropped NBC and became an ABC affiliate. Naturally, this switch did not sit well with WTHR TV. VideoIndiana, the Columbus, Ohio, owner of WTHR TV, decided to sue ABC and the McGraw-Hill Broadcasting Co., Inc. (owner of WRTV). The McGraw-Hill TV station in San Diego had also switched from NBC to ABC. WTHR charged that this constituted a "conspiracy" and "unreasonable restraint of trade." The suit sought $33 million in damages.

The suit was dismissed, and the network switch took place on June 1, 1979. Chris Duffy, the colorful general manager of WTHR, used his imagination as a former station promotion manager in Chicago to seek a last ounce of revenge. At 1:40 a.m. on June 1, 1979, he scheduled a program called *Execution of Affiliates Cup*. Just before the "Star Spangled Banner," Duffy took the cup the station had received from ABC signifying 20 years of network affiliation and placed it on a stand. He then took careful aim with a rifle, and shot the cup full of holes.

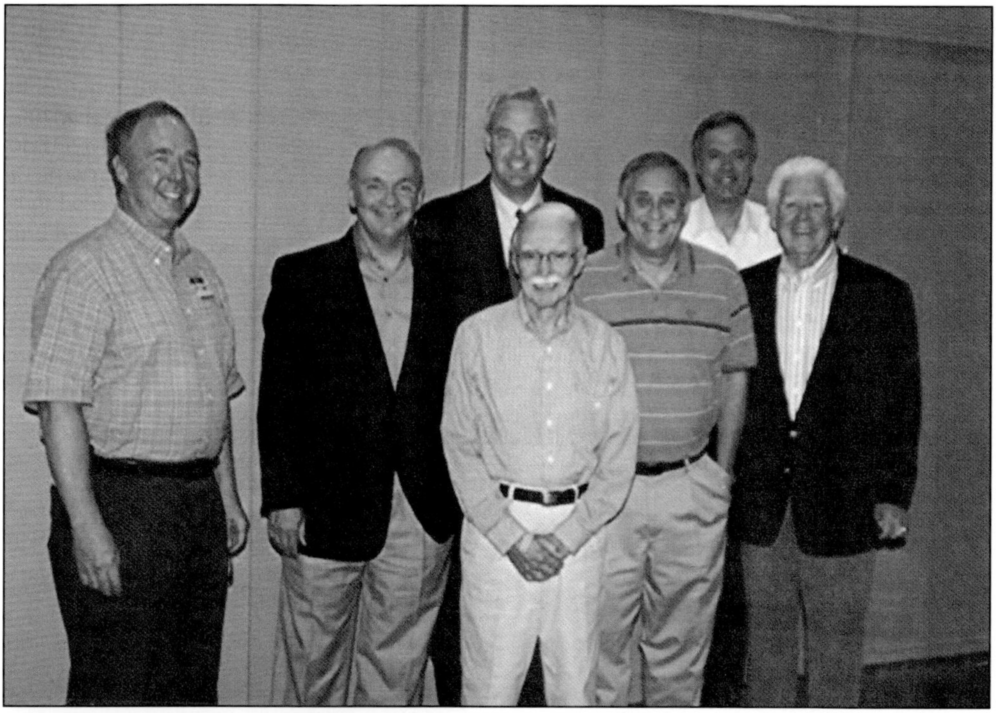

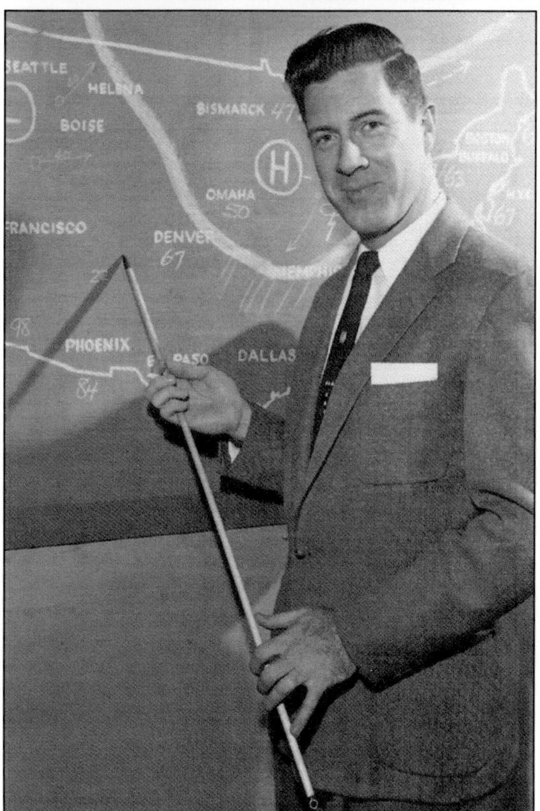

In 2010, the Indiana Broadcast Pioneers hosted a luncheon for television weather pioneers. From left to right are David James, Chuck Lofton, Kevin Gregory, Stan Wood, Bob McClain, Randy Ollis, and Bob Gregory.

In the early days of television, there were not enough legitimate weathercasters on the payroll to cover all the time slots for weather; announcers or other talent were called upon to do a weathercast now and then. Here is Joe Leamon at WISH TV as a fill-in weatherman.

Frank Prater of WFBM TV did a lot of jobs, not the least of which was taking on the role of weatherman. Note that he is outdoors and has several weather balloons.

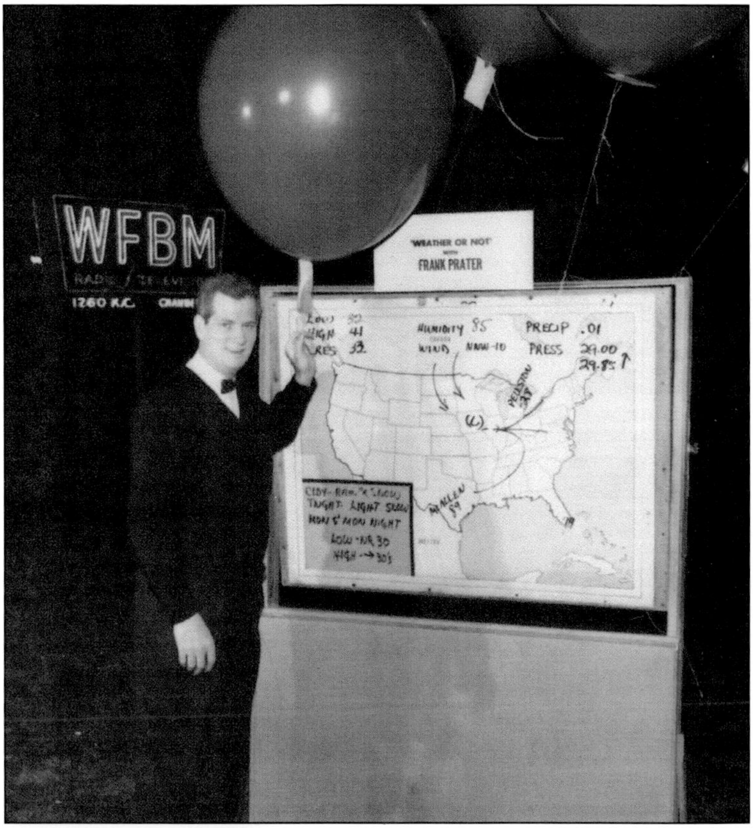

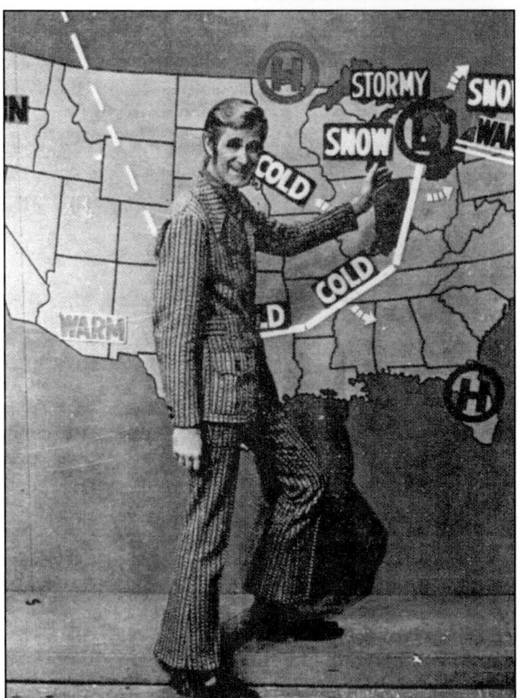

Jimmy Mack at WLWI TV was an announcer and the host of *Bandstand 13*. Here, he is properly attired for his role as a weatherman.

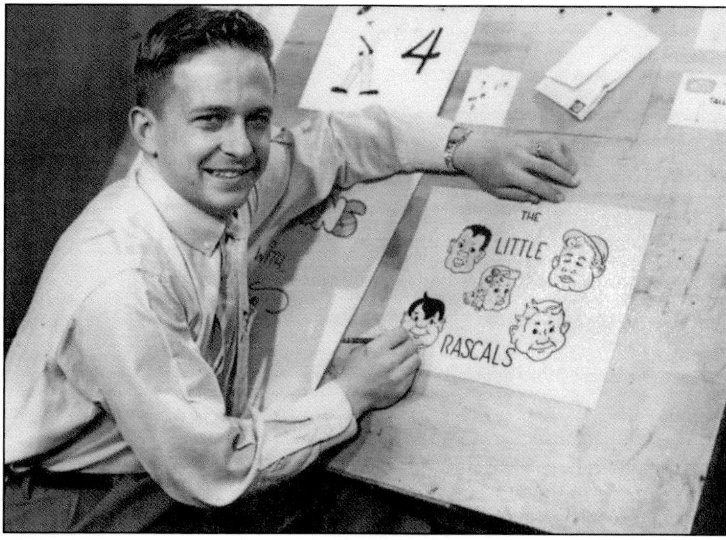

Les Satterthwaite was a cartoonist/weatherman. He was a regular weatherman at WTTV, and he was also the host of *The Little Rascals*.

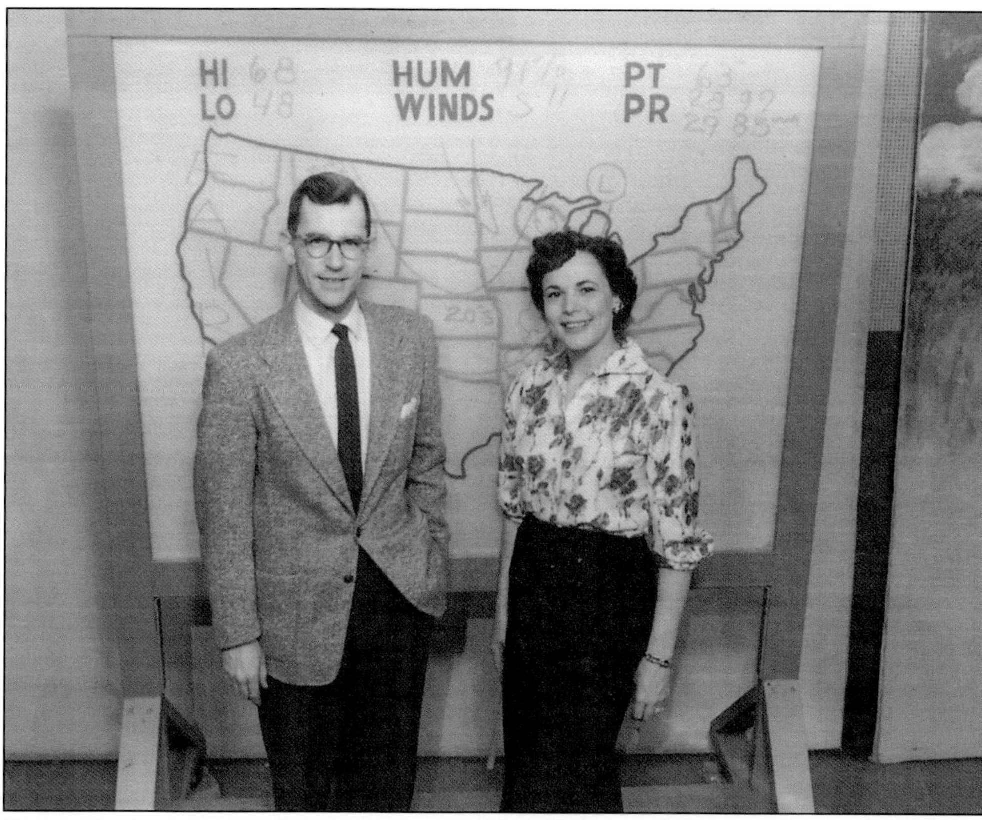

Frequently, the commercial announcer appeared with the weatherperson. Here is Bill DeWitt, who did the Motorist Mutual commercials for Kay Field's weathercasts.

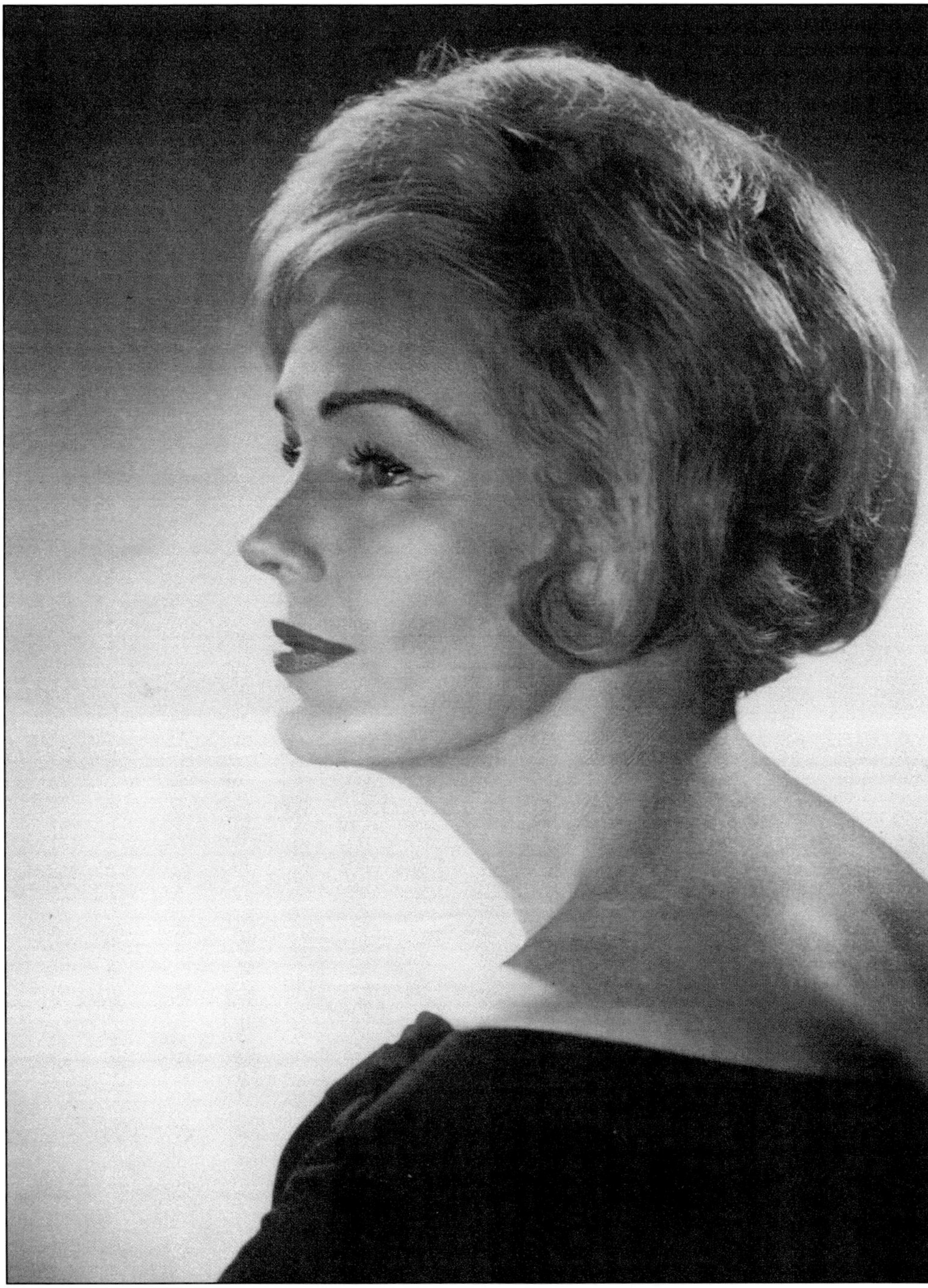

Frances Farmer was a stage and movie actress. She was an accomplished actress but was troubled by alcoholism and erratic behavior. Her movie career was over and she was attempting an acting comeback when she began touring in a production of *The Chalk Garden*. It was at Avondale Playhouse in Indianapolis that she received an offer to host movies on WFBM TV. Her husband at that time was Lee Mikesell, who was from Indianapolis, and this was an opportunity to have steady employment.

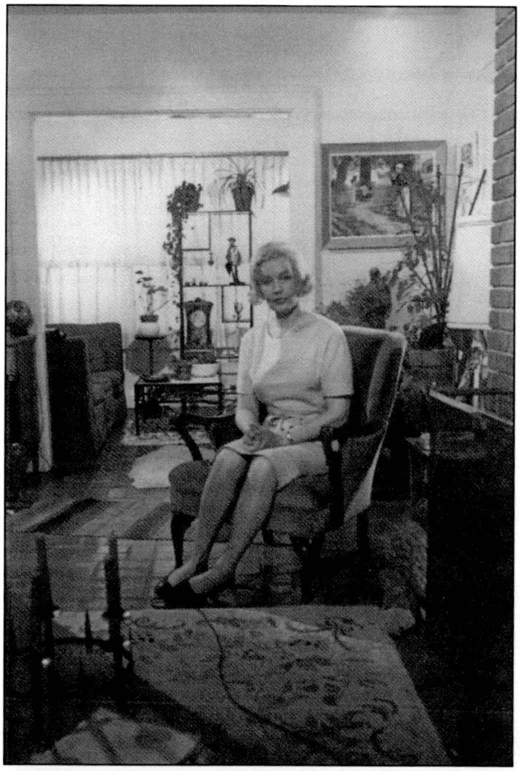

Her television program was entitled *Frances Farmer Presents*, and it was not long before she became an audience favorite. Farmer talked of her days in Hollywood and, frequently, had celebrity guests on the show.

Eventually, Frances Farmer's erratic behavior returned, and she was fired, rehired, and fired again. Her tenure as a movie host ran from 1958 to 1964. After she left the television show, she continued to live in the Indianapolis area until her death in 1970. In 1982, Jessica Lange played Farmer in the film *Frances*. Susan Blakely played her in a 1983 TV production entitled *Will There Really Be a Morning?*, which was the title of her autobiography.

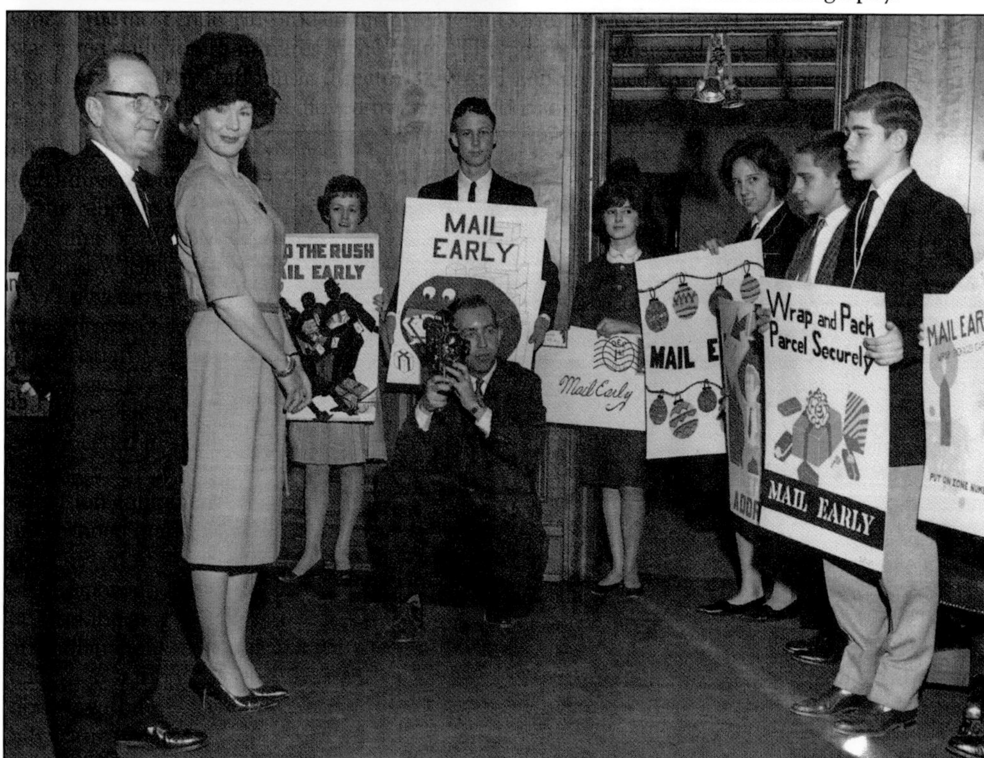

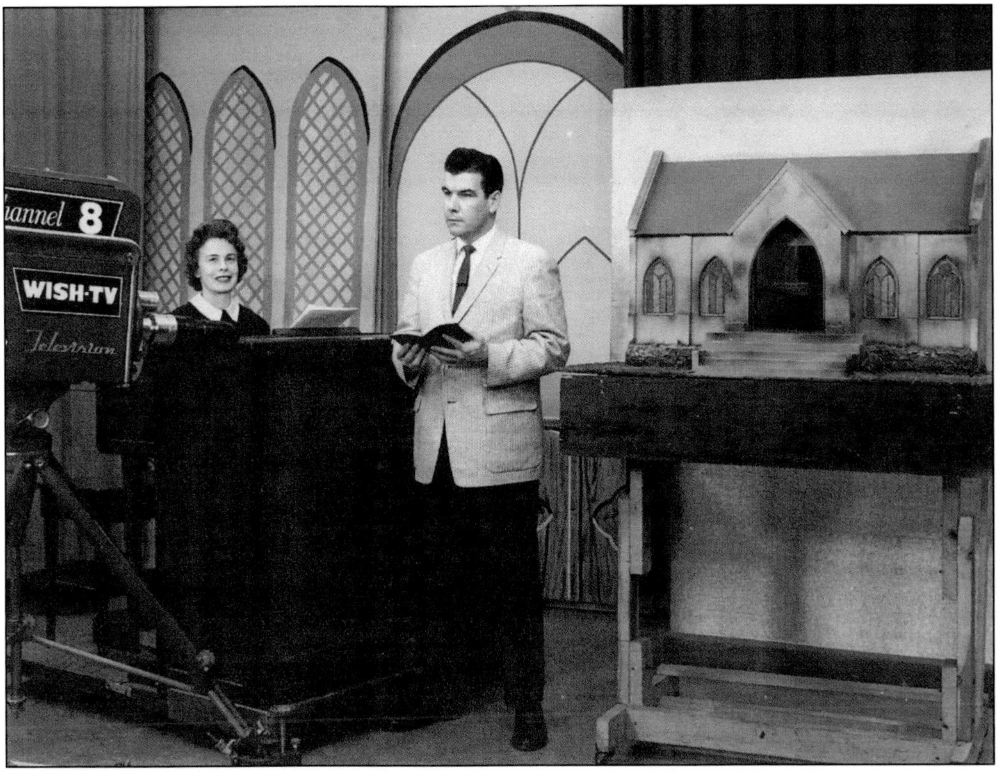

One of the most endearing and enduring program series was WISH TV's *Chapel Door.* The program was aired daily for 15 minutes at 7:00 a.m. The perennial host was Rev. Russ Blowers. Virginia Byrd served as the staff music director at WISH and provided the music. Note the mock-up on the right where the camera moved in and the doors opened.

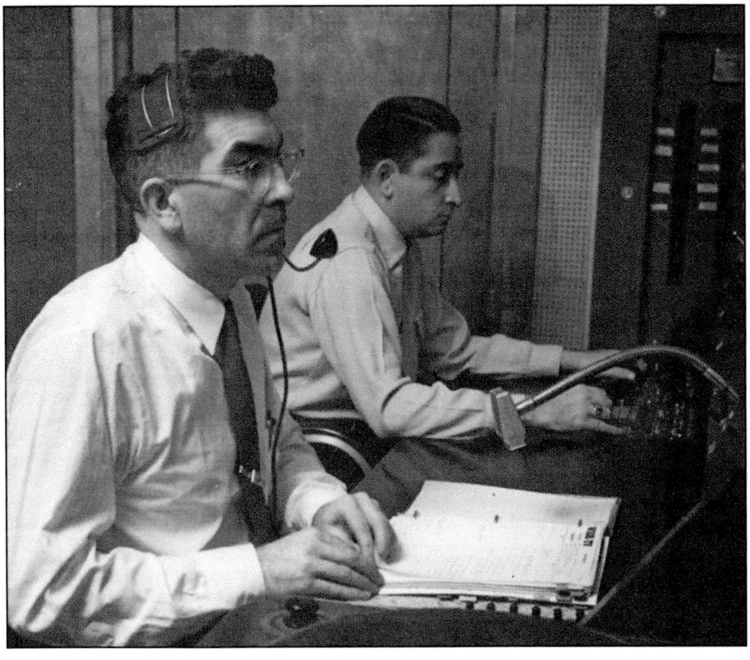

Dr. Al Edyvean (left) directed most of the religious programs on WFBM TV and WISH TV. Edyvean was a professor at Christian Theological Seminary. He convinced the seminary to buy television equipment and trained many students, most of whom wound up working at an Indianapolis television station. Shown with Edyvean is engineer Jerry Cooper.

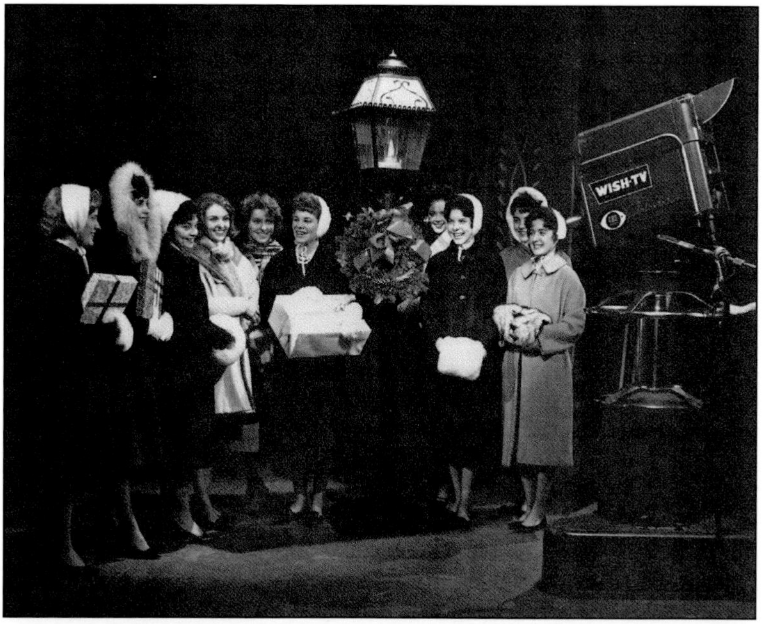

Religious specials were aired on all of the stations in the market, especially at Christmastime.

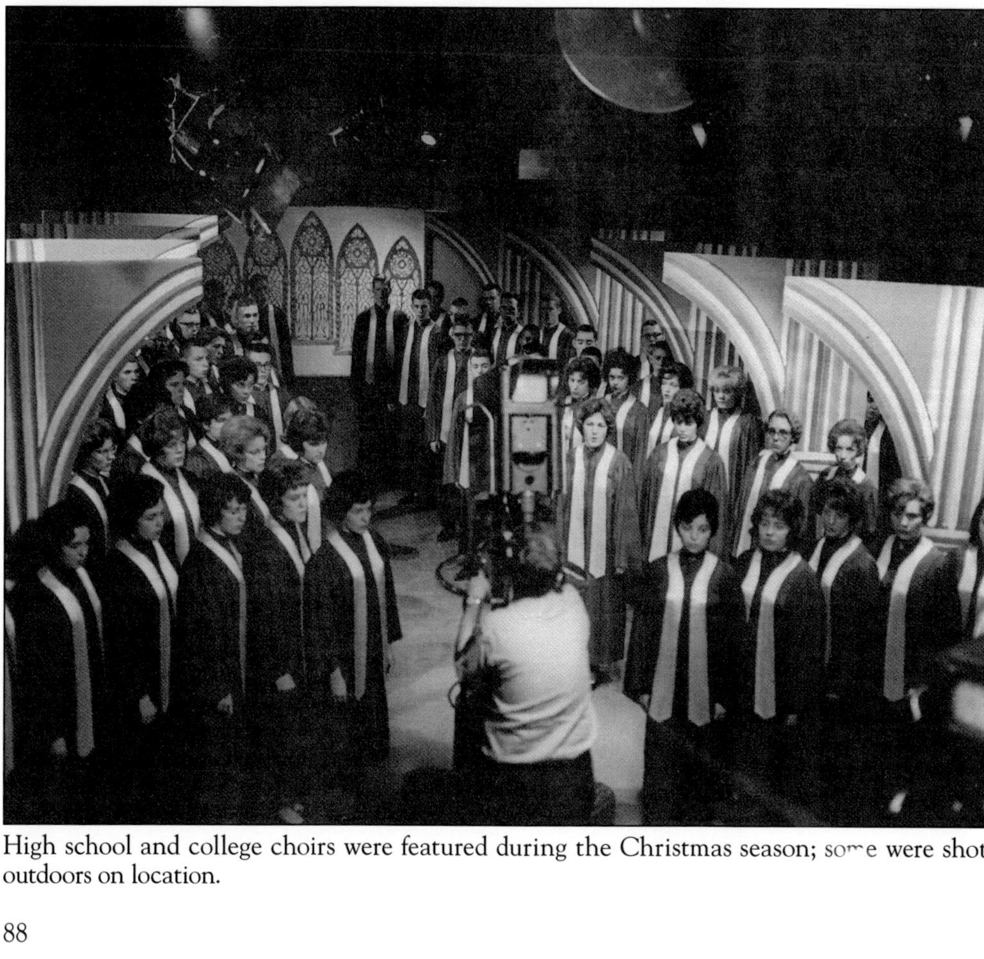

High school and college choirs were featured during the Christmas season; some were shot outdoors on location.

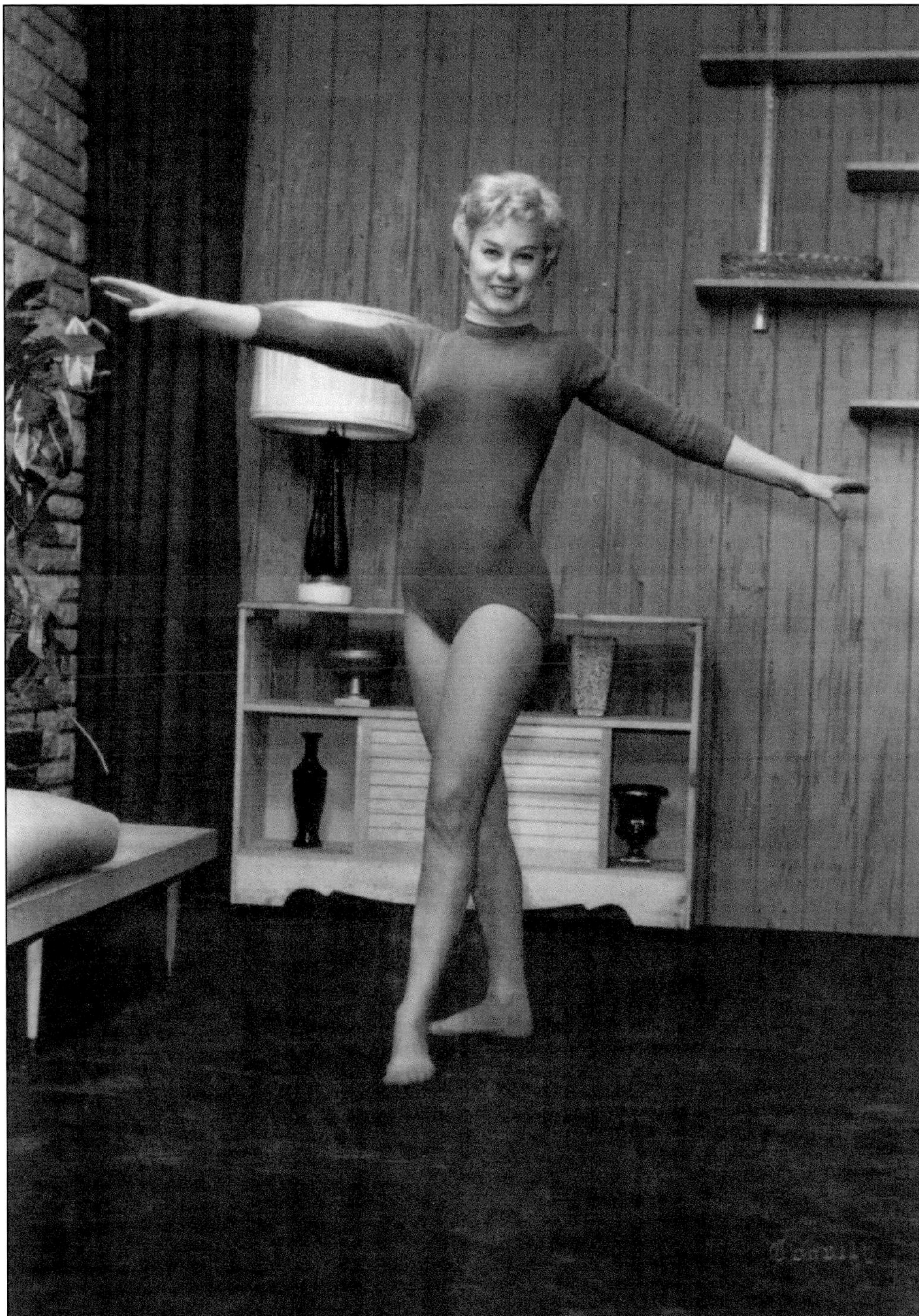

Not many local stations can originate a program that becomes nationally syndicated. In 1960, WISH TV discovered an exercise girl who did go national. Debbie Drake came to Indianapolis from Dayton, Ohio, where she had a weekend exercise show. WISH TV offered her a daily 15-minute show in the morning.

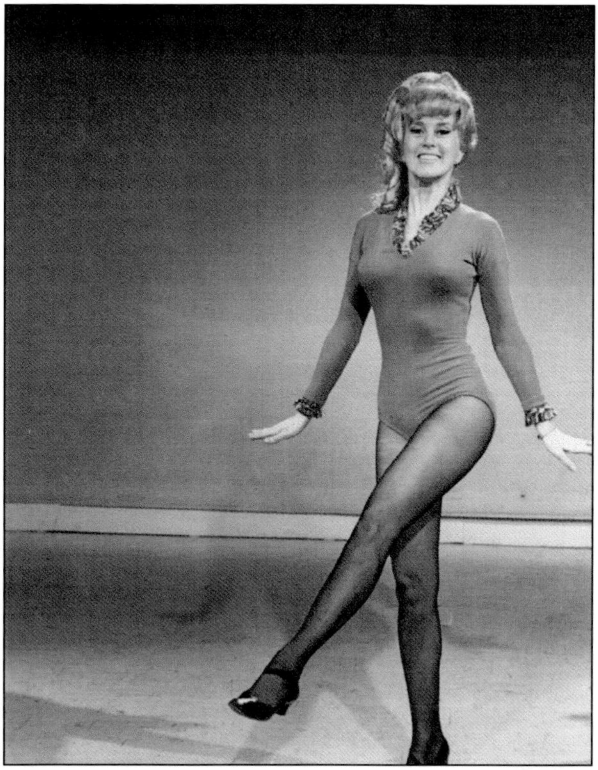

It was not long before it became evident that this was a trailblazing show. Ratings were terrific, and sponsors were signing up. WISH TV's parent company, Corinthian Broadcasting, decided to syndicate the show, and the syndicated series was produced in the Indianapolis studios.

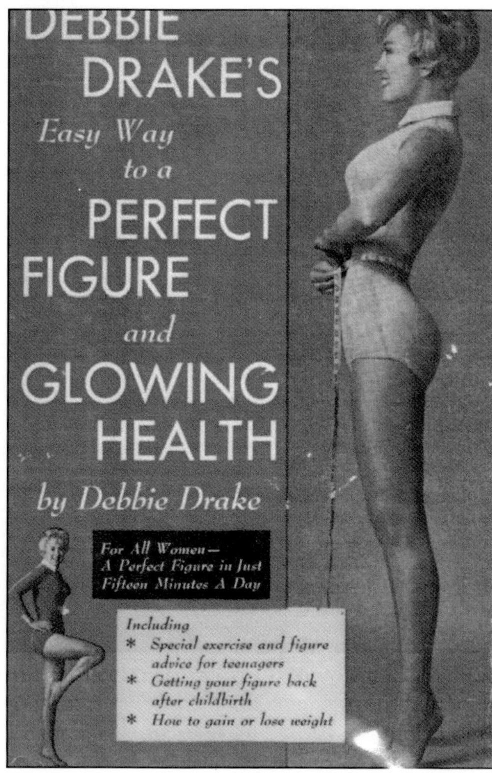

The Debbie Drake Show was sold in over 100 markets in the United States and went into markets in Europe and South America. In 1961, Drake published a book and began making appearances on *The Tonight Show* with Johnny Carson, *The David Frost Show*, *What's My Line?*, and many others. She made two record albums.

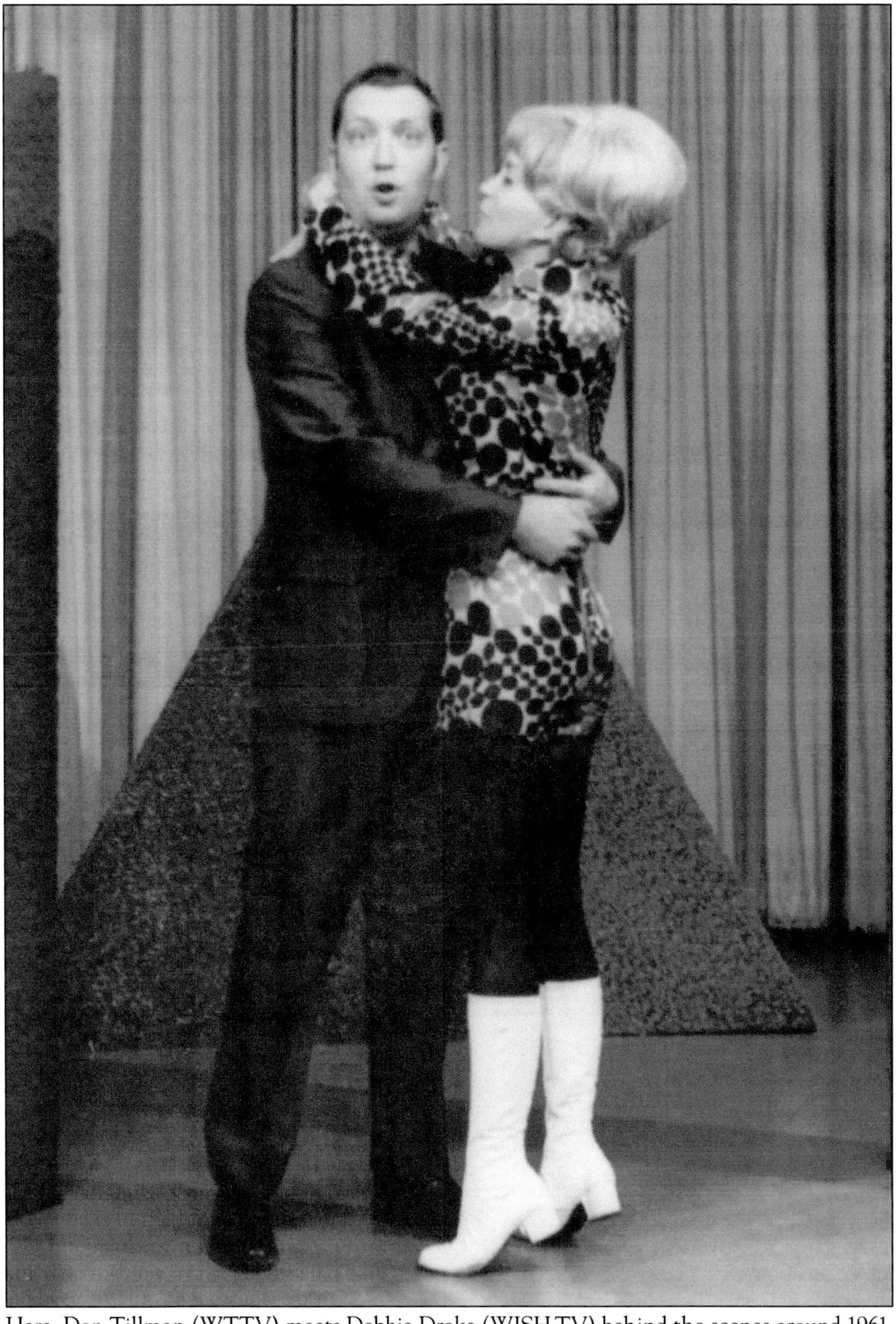

Here, Don Tillman (WTTV) meets Debbie Drake (WISH-TV) behind the scenes around 1961.

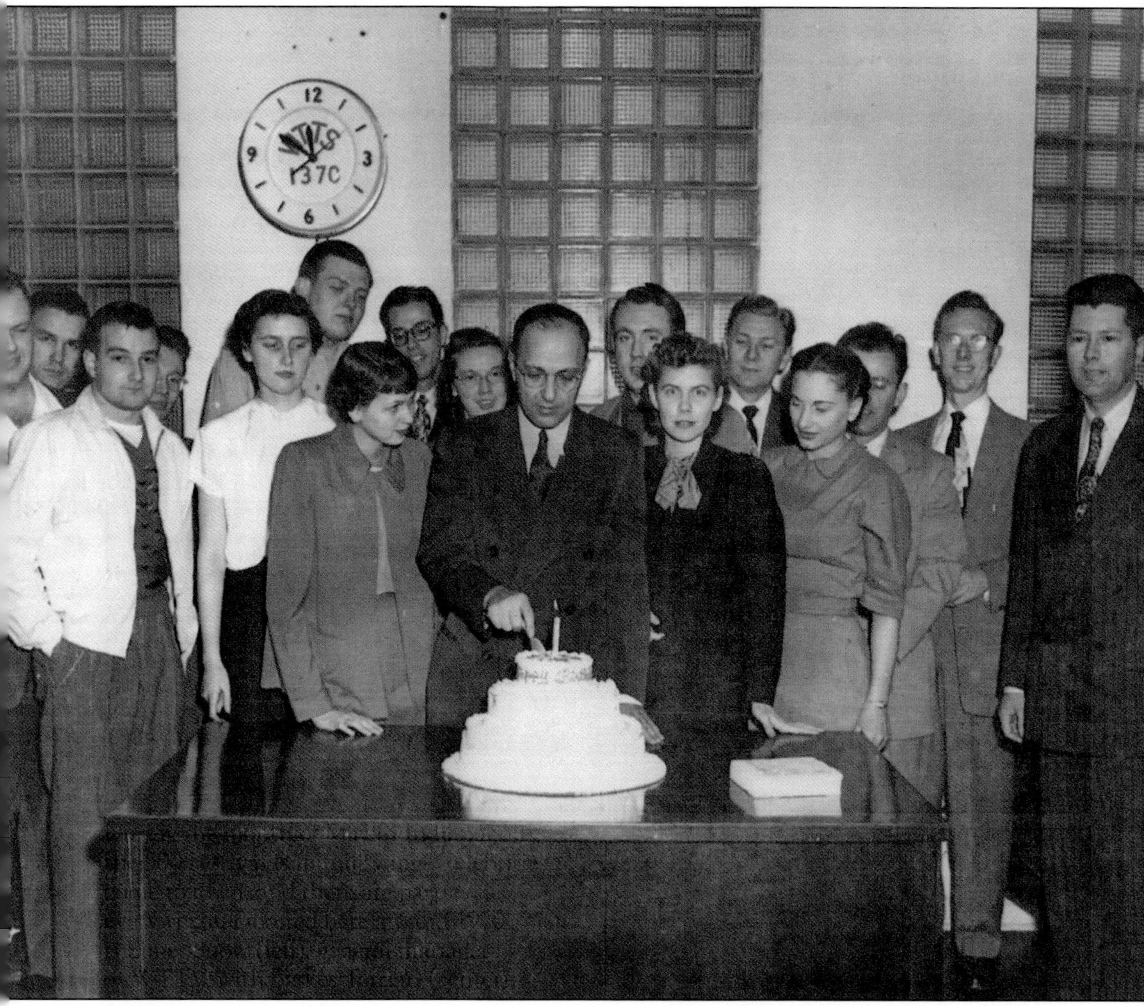

WTTV waged a continuing battle to improve its viewership in the Indianapolis area. However, when cable emerged, the question of tower location was not nearly as important as it once was. Ready to cut the cake is Sarkes Tarzian. To the left of Tarzian is Sue Bartlett, and on the far right is Glenn Van Horn, who was the first general manager of WTTS and WTTV.

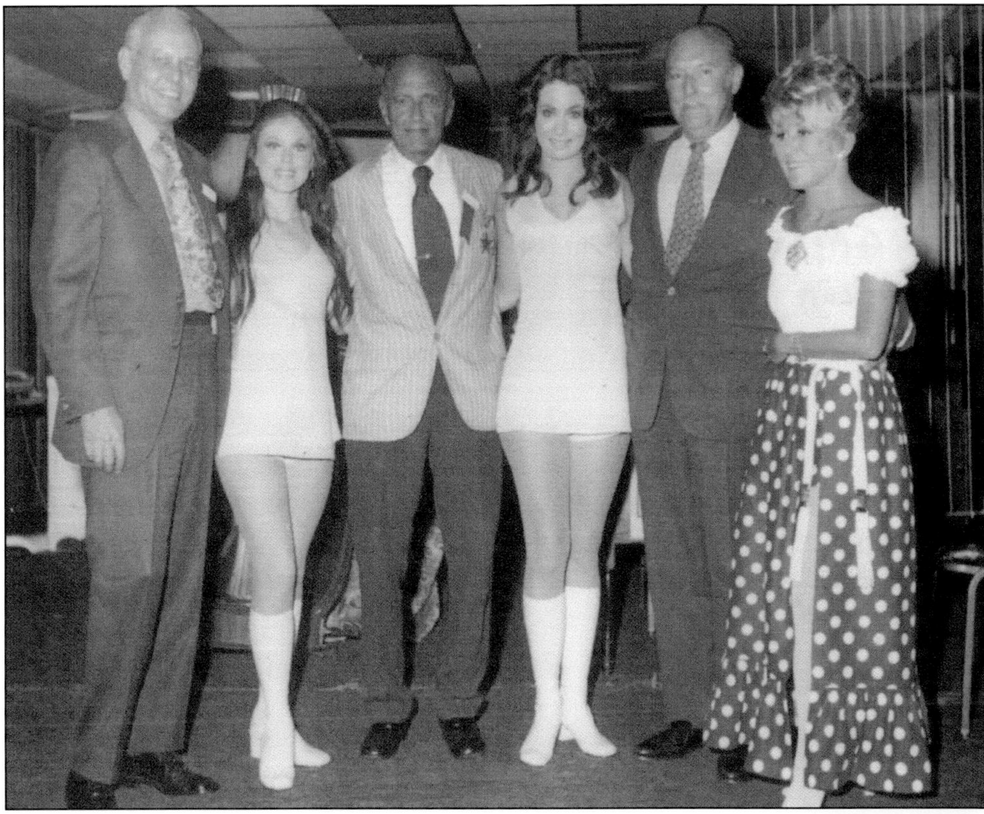

During the month of May in Indianapolis, the 500 Festival is in full swing. Celebrities come to town, and there are parties everywhere. Pictured here around 1970 is Elmer Snow (left), general manager of WTTV, with Sarkes Tarzian (center) and actor Gale Gordon. Columnist Rona Barrett is to the far right. The two girls in the shot are "Gold Diggers" from Dean Martin's TV show.

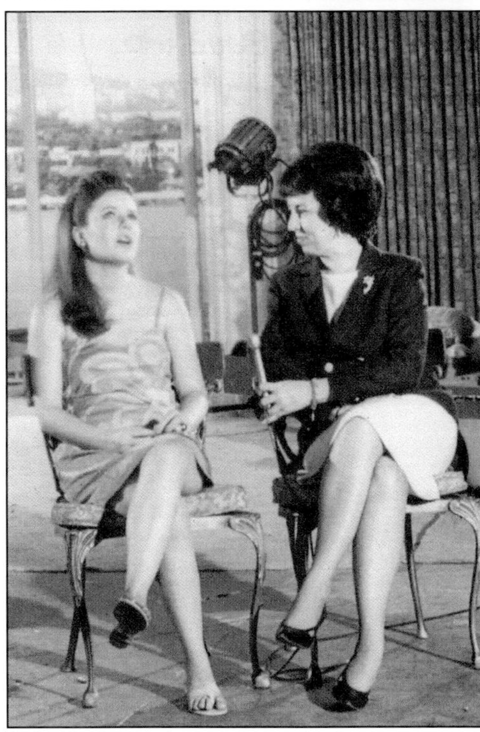

One of the more popular talk/interview shows was *The Billie Boucher Show* on WTTV. Here, Boucher (right) is interviewing actress Patty Duke. Boucher's show ran from 1962 to 1968.

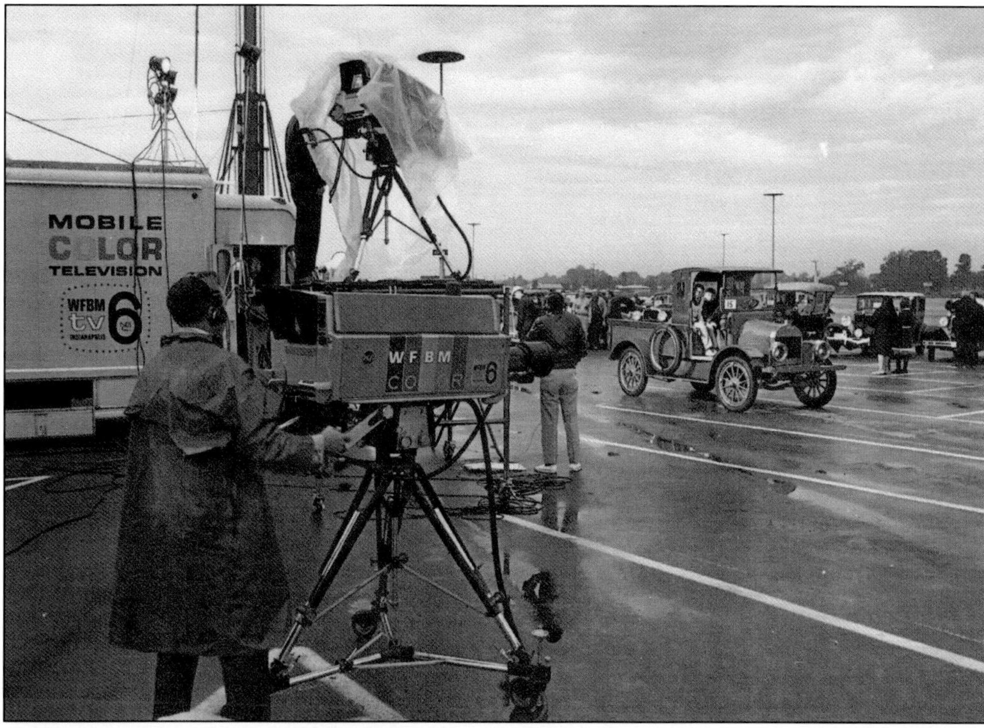

A popular special event created by WFBM TV was its annual *Antique Auto Tour.* Most of the station's personalities participated under the guidance of Tom Carnegie and promotion director K.C. Strange. The station selected 150 autos from over 300 applicants. Starting at the Indiana State Fairgrounds, the autos made their way through Noblesville, Tipton, Alexandria, and Anderson.

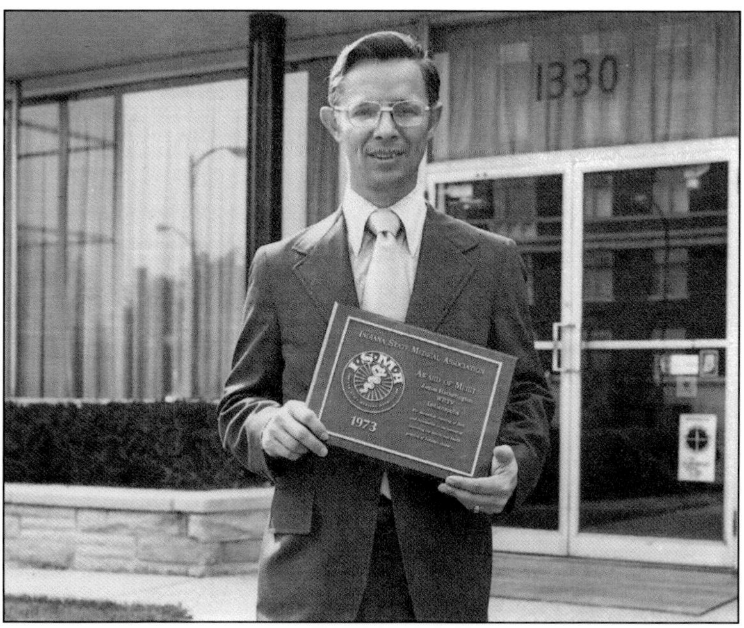

WFBM TV was very active in producing award-winning local documentaries in the mid-1960s and early 1970s. The primary responsibility for these documentaries was given to editorial and documentary editor Jim Hetherington. Hetherington was a longtime journalist and public relations executive. He worked for the *Indianapolis Times* before moving to WFBM in 1963. He left WFBM TV in 1974

Steve Scott was manager of public affairs for the WFBM stations in the late 1960s. Here, he receives a Casper Award from the Community Service Council of Indianapolis.

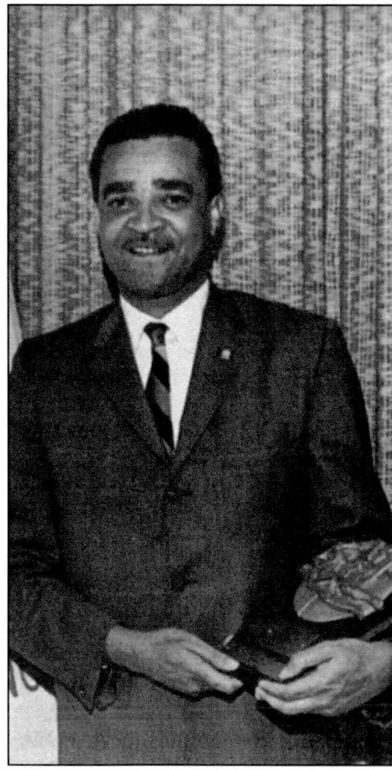

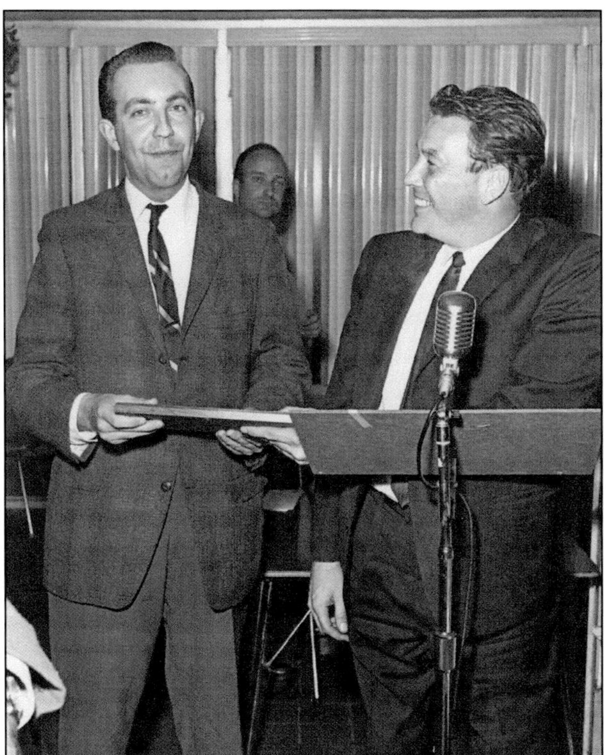

WFBM TV photographer Dick Baldwin (left) receives an award at Indiana University. Presenting the award is Dick Yoakam. Yoakam was a venerated professor at Indiana University, where he taught for 32 years. He counted among his students Dick Enberg and former CBS White House correspondent Phil Jones. He taught both journalism and telecommunications, retiring in 1989.

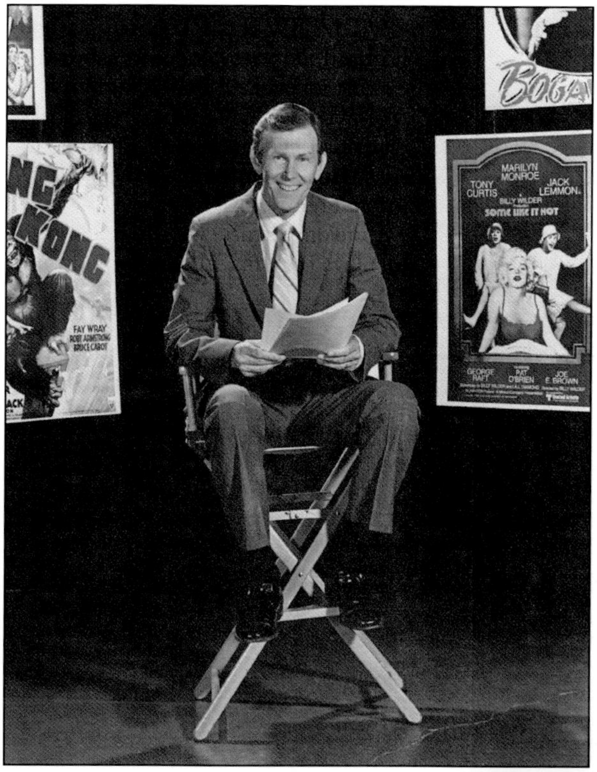

In 1971, Dave Smith, who was program manager of WISH TV, started hosting a thematic series of movies mostly from the 1930s and 1940s. It was entitled *When Movies Were Movies*. It ran for 10 years.

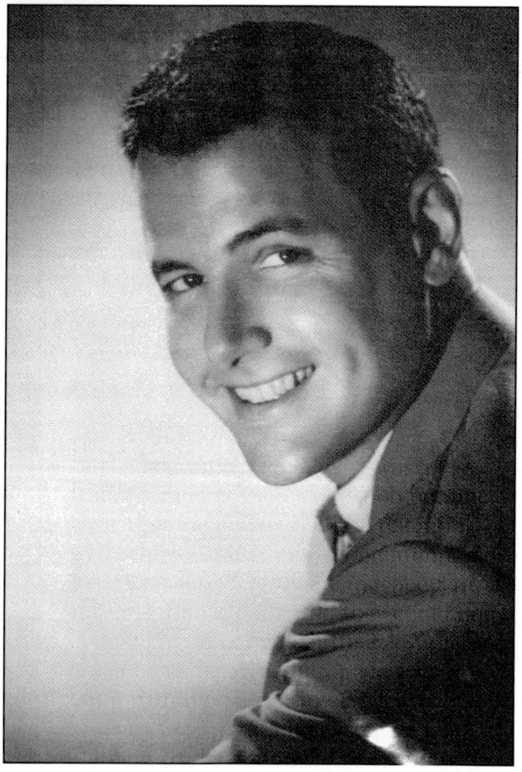

Bernie Herman, who was a disc jokey on WFBM Radio, moved over to television and took the place of Frances Farmer as host of the WFBM TV afternoon movies.

Reid Duffy toiled in the WFBM TV newsroom for many years, but his droll sense of humor and his love for the culinary arts led him to create *Duffy's Diner.* Duffy traveled throughout the WFBM TV coverage area visiting a variety of restaurants and then reported his opinion of the cuisine, ambiance, and whatever else he cared to evaluate.

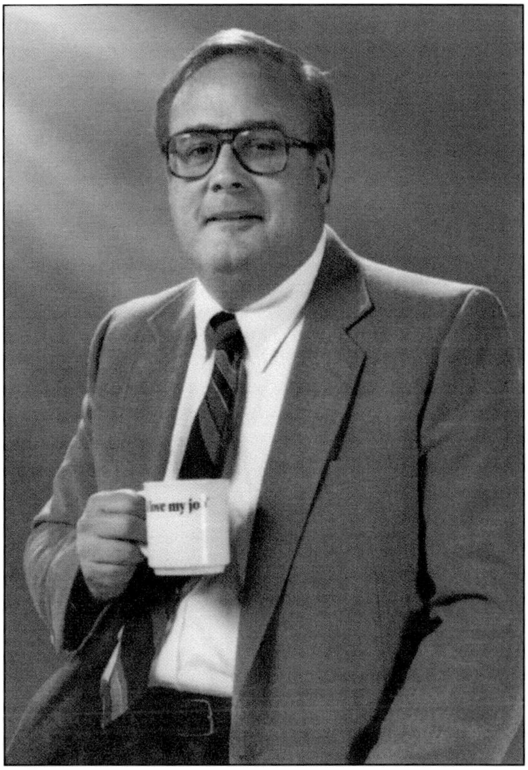

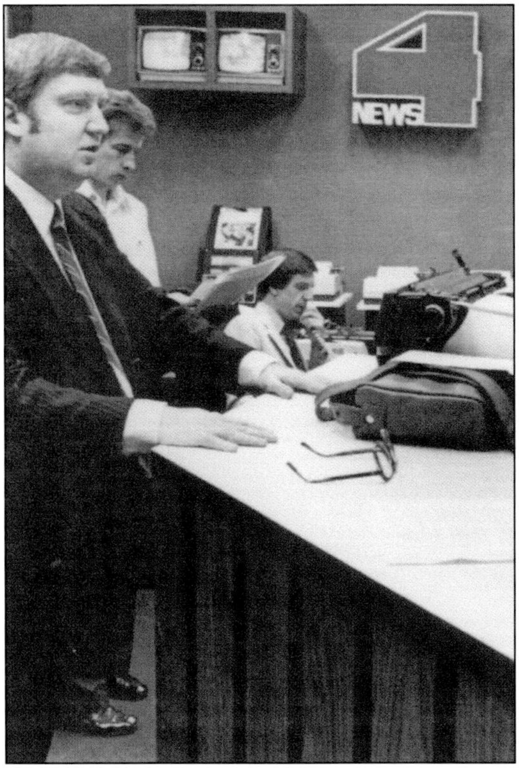

In the beginning, local news was a loss leader for many stations. It soon became clear that local news had the potential to be the most lucrative programming on a station's schedule. Some independent stations tried it but found they could not compete with network affiliates. WTTV had a news department for a few years, but it dwindled down to very few people. Shown is programming executive Don Tillman in the channel 4 newsroom.

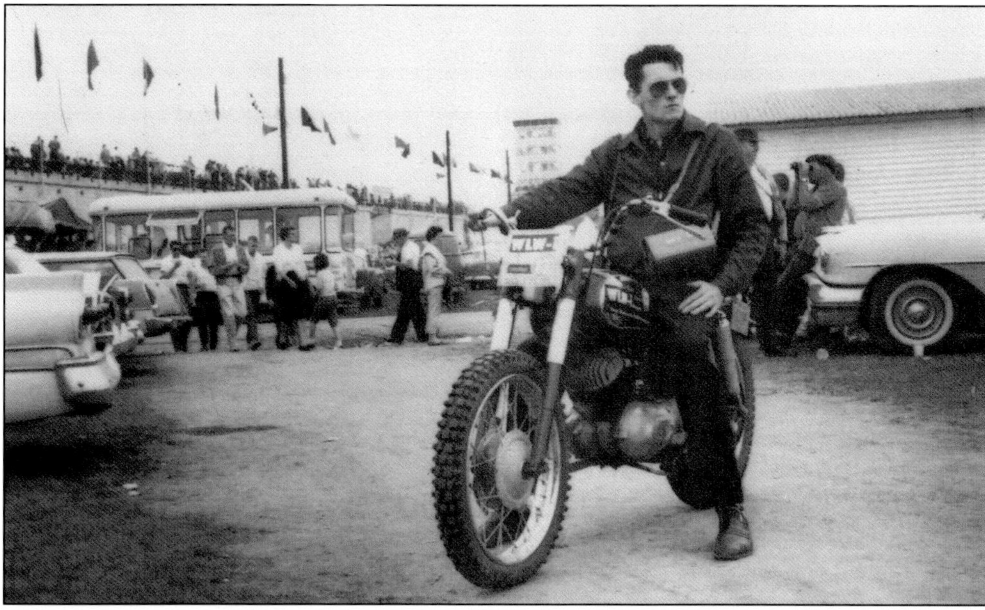

A handicap in the days before videotape was the necessity to shoot, process, and edit news film. Here is Dan O'Leary of WLWI TV on his motorcycle at the Indianapolis Motor Speedway. O'Leary took the film from the photographers and raced back to the TV station to get it processed as soon as possible. O'Leary also was a courier for the station.

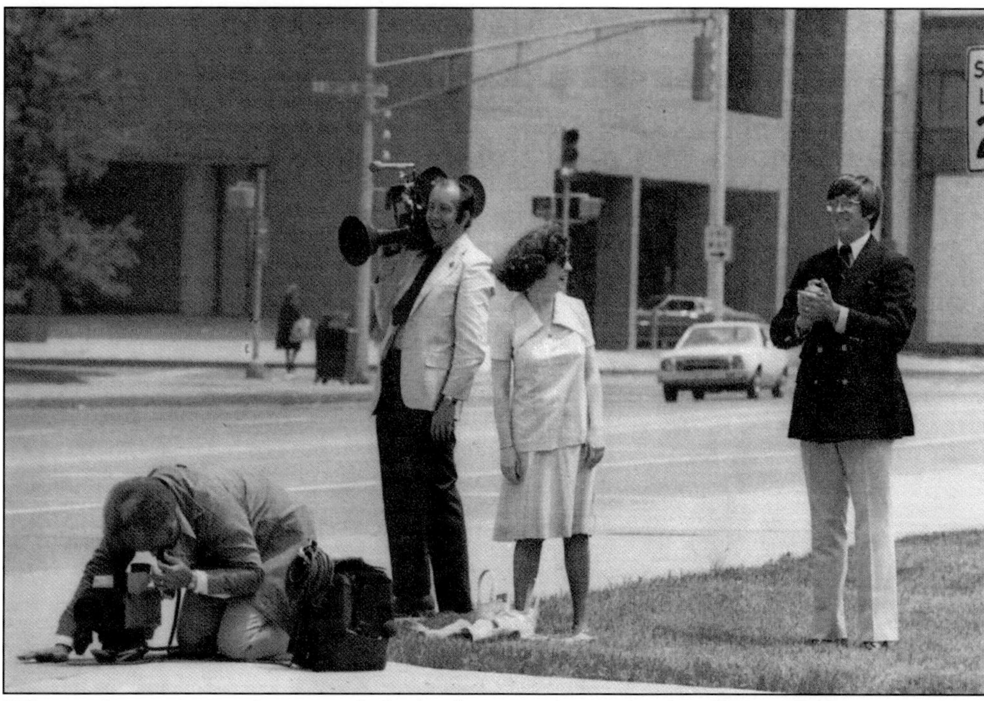

When videotape arrived, it provided a big boost with timelier handling of news stories. Here are two stations shooting the same story with their portable tape cameras. From left to right are Paul Brady, WISH TV; Dick Baldwin, WFBM TV; Linda Lupear, WFBM TV; and Bruce Childs, WISH TV.

Videotape also made it easier to do remote telecasts of special events. In 1961, a WISH TV crew taped a special from Monument Circle. From left to right are (first row) Steve Rickel, Ron Stroup, and Paul McAllister; (second row) Stan Wood, Harry Heuston, Ron Bowler, Al Szaley, Ed Lee, and Lynn Edwards.

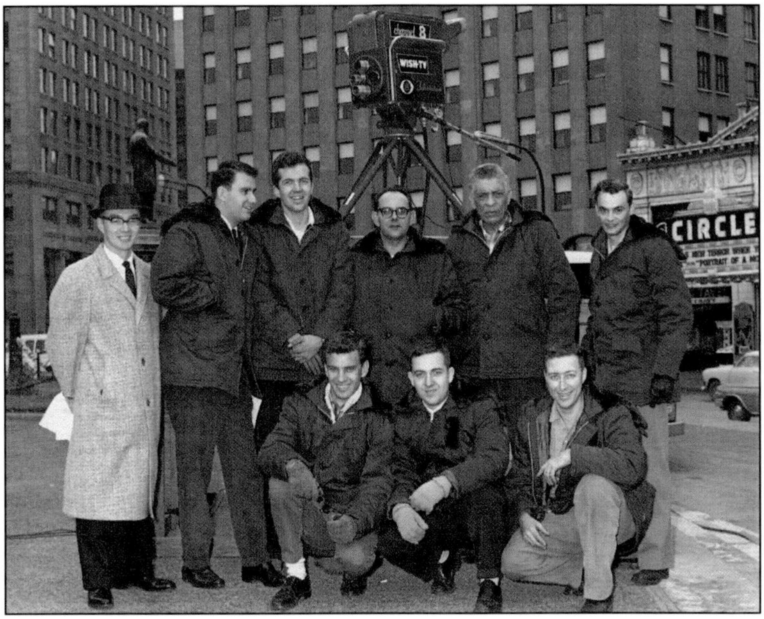

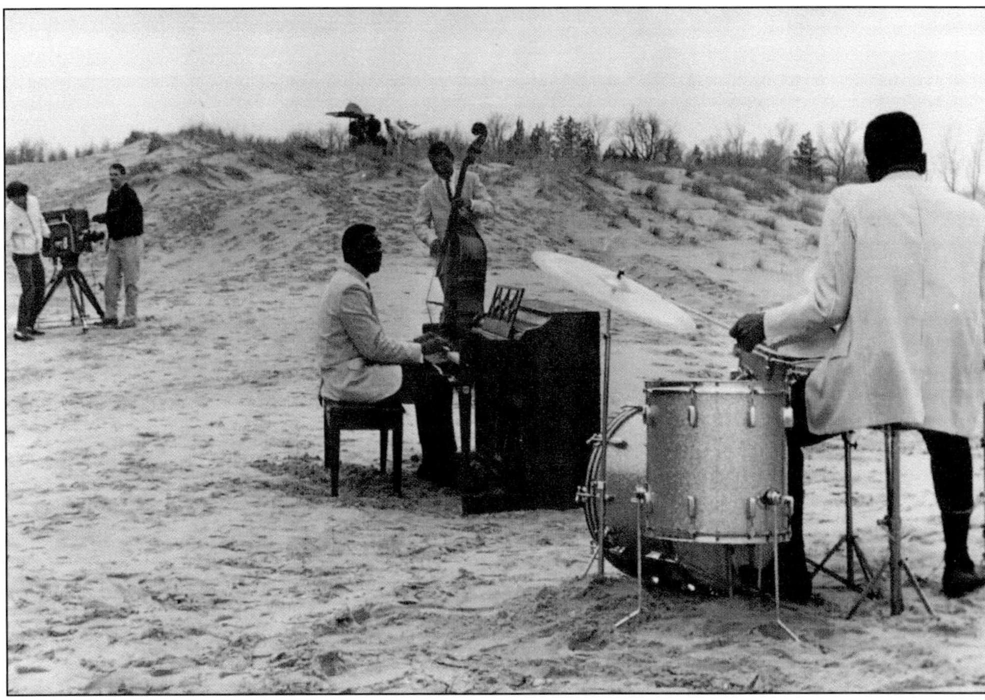

In 1965, WISH TV produced an hour-long special called *Campus Talent*. It was shot and videotaped at various locations throughout the state, many on college campuses. Shown is a musical group on location at the Dunes State Park on Lake Michigan. The program became an annual event for the station.

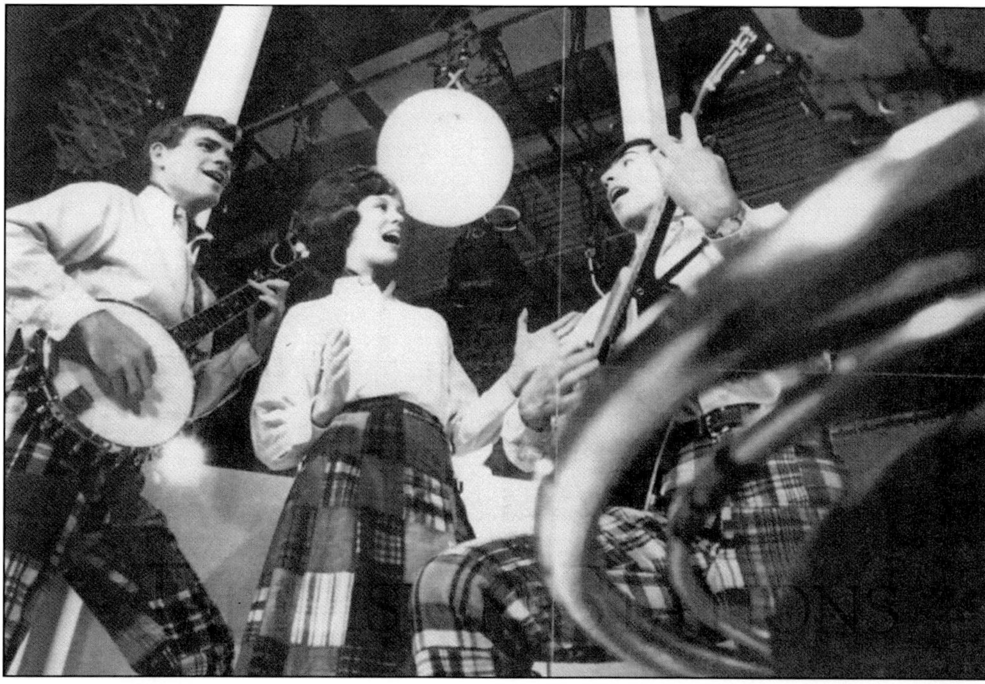

Each year, a winner was declared and sent to New York City to audition before professional scouts and agents. Shown are the Scooters from DePauw University, the winner of the first *Campus Talent* competition.

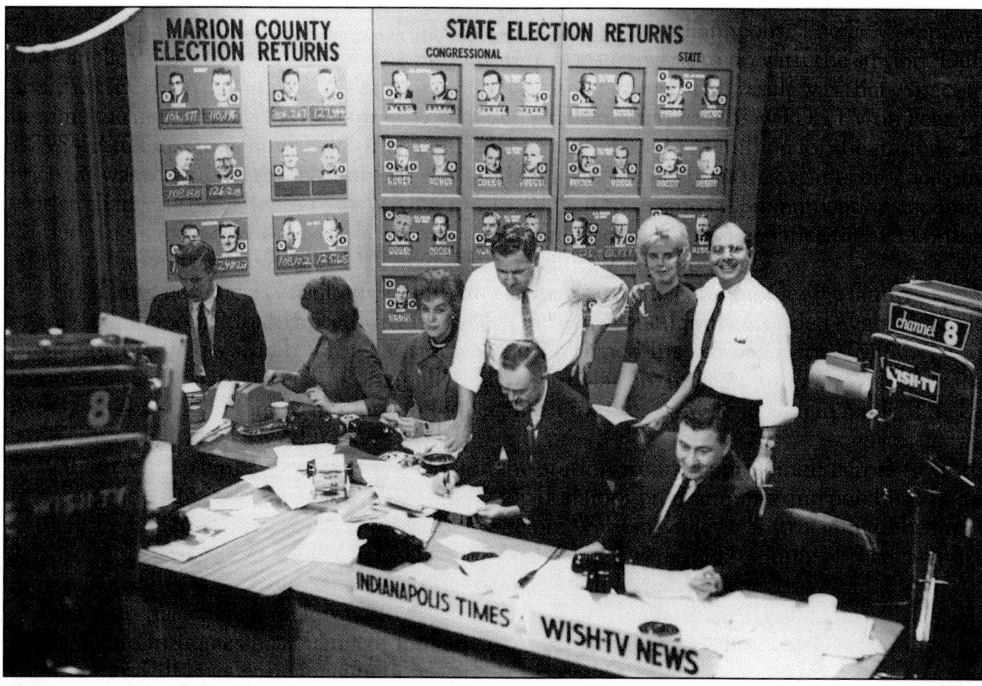

Shown is some of the election coverage in the 1960s at WISH TV. From left to right are Dave Smith, Judy Harrison, Marifrances Curtis, Ray Sparenberg, Richard Hickox, Mary Jo Peck, Joe Mingioli, and Irving Leibowitz of the *Indianapolis Times*.

Ten

The Last Television Stations

After six years running WTHR TV, Chris Duffy abruptly resigned to join start-up Anacomp, Inc., as vice president in charge of communications and advertising. In 1978, Melvin and Herb Simon, shopping-mall owners, applied for UHF channel 59 in Indianapolis. The FCC approved their application in 1983; however, the Simons were not sure what to do with the station. Duffy and the CEO of Anacomp, Ron Palamara, met with the Simons. The result was that Anacomp became the majority owner of channel 59. The station went on the air as WPDS TV on February 1, 1984. After just nine months, Outlet Communications offered $22 million for the station. The price tag was reportedly the most money ever paid for a TV station in its first year of broadcasting. Outlet changed the call letters to WXIN TV. In 1995, Renaissance Communications acquired Outlet in a merger valued at $360 million. In 1996, Tribune bought Renaissance for $1.13 billion, adding 6 stations to its group of 10. Tribune also bought WTTV in 2002.

Chris Duffy became a consultant and began to work with the Hulman-George family, owners of the Indianapolis Motor Speedway. In 1986, Duffy and the family founded Wabash Valley Broadcasting. In his search for a TV station, Duffy turned his attention to WMCC TV on channel 23, a TV station that was originally assigned to Marion, Indiana. In 1995, Wabash Valley purchased the station for $10 million, and it became a fixture on cable in Indianapolis. The call letters were changed to WNDY TV.

Wabash Valley built large headquarters for the station in a shopping mall across from the Indianapolis Motor Speedway and began to pay top dollar for programming and purchased state-of-the-art production equipment. It was not long before Wabash Valley was struggling. The family decided to sell it. In 1998, all of the assets of Wabash Valley were sold to Emmis Broadcasting of Indianapolis. In 2005, most of these assets were sold to LIN Media, owner of WISH TV.

Television has come a long way since Jerry Smith built his homemade TV receiver in 1938. That picture in homes has changed dramatically over the years, and rest assured, changes are coming that will boggle the imagination.

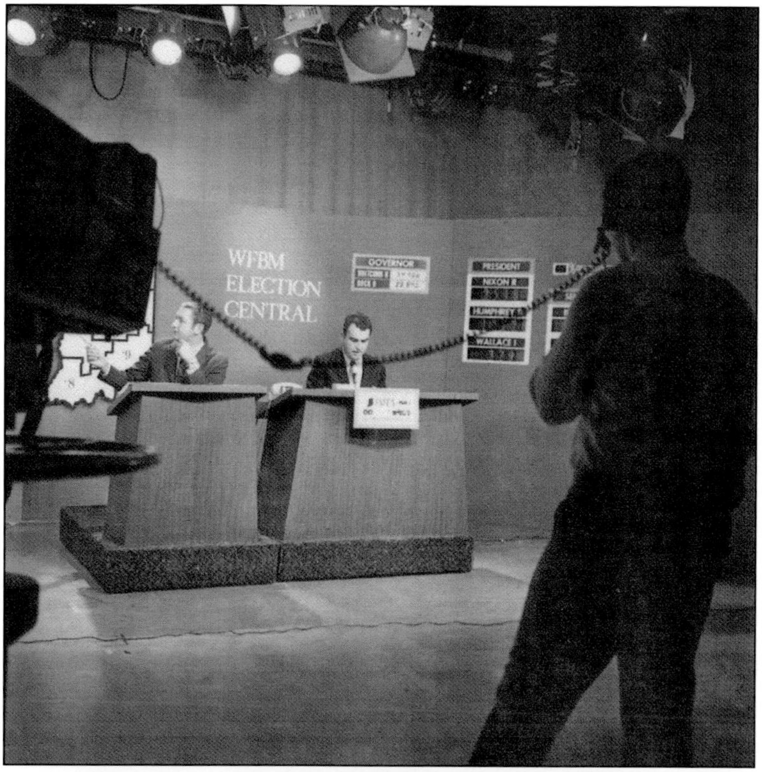

Howard Caldwell is covering a 1967 mayoral election from the WFBM TV studios.

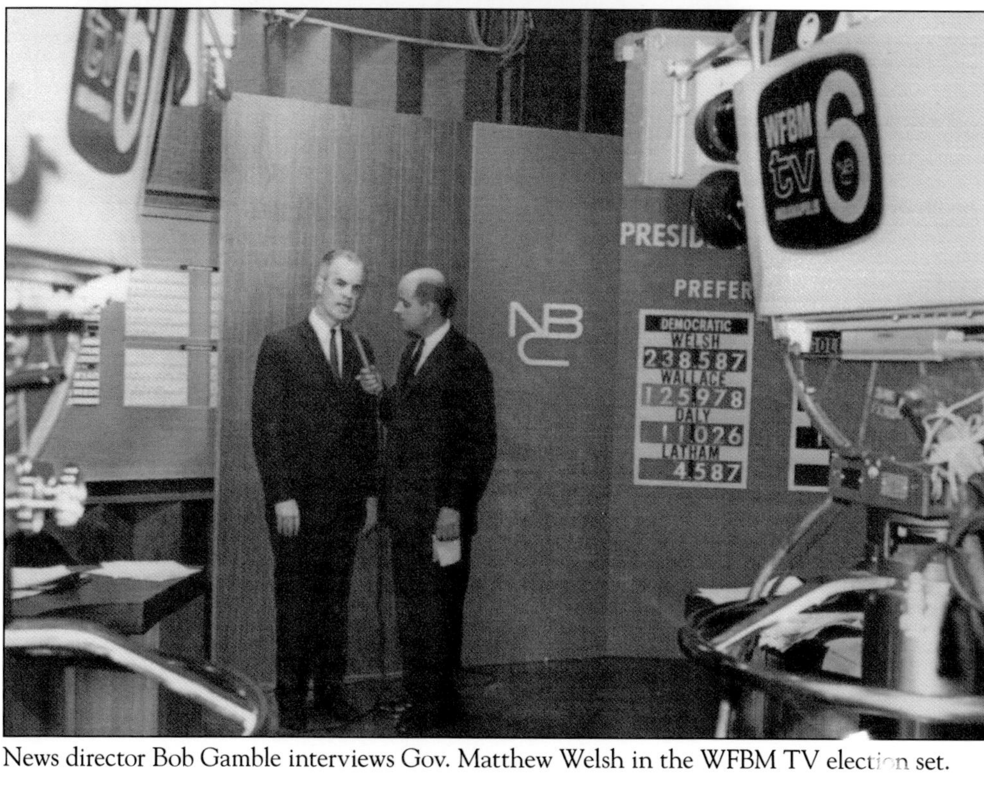

News director Bob Gamble interviews Gov. Matthew Welsh in the WFBM TV election set.

Chuck Marlowe, long-time sports director for WTTV, is probably best known for his 29-year tenure as Bob Knight's television partner. Knight had a weekly TV show on WTTV during the basketball season. Highlights of the game were shown with Knight providing the commentary. Marlowe managed to keep things under control during this long television partnership. He and Knight are still the best of friends.

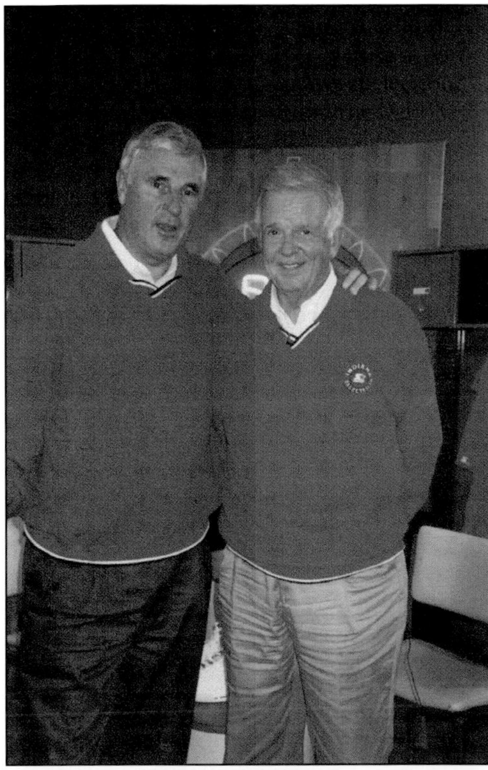

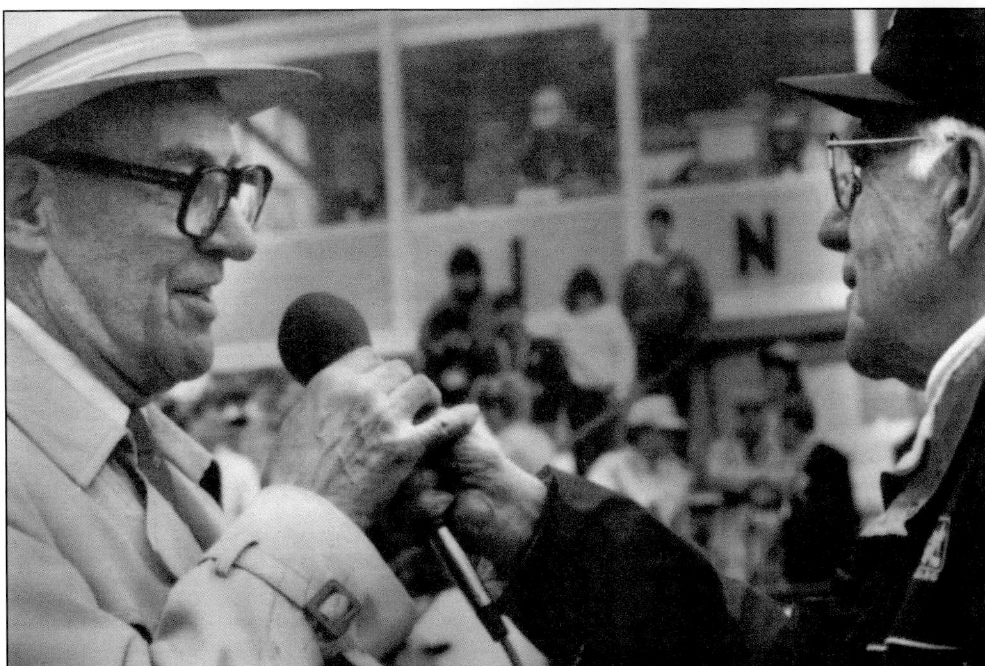

Earl Townsend Jr. (left) and Tom Carnegie are at Indianapolis Motor Speedway in 1989 for a ceremony honoring the 40th anniversary of the first televised Indianapolis 500 Mile Race. Townsend called the play-by-play of that race in 1949. (*Indianapolis Star* file photograph.)

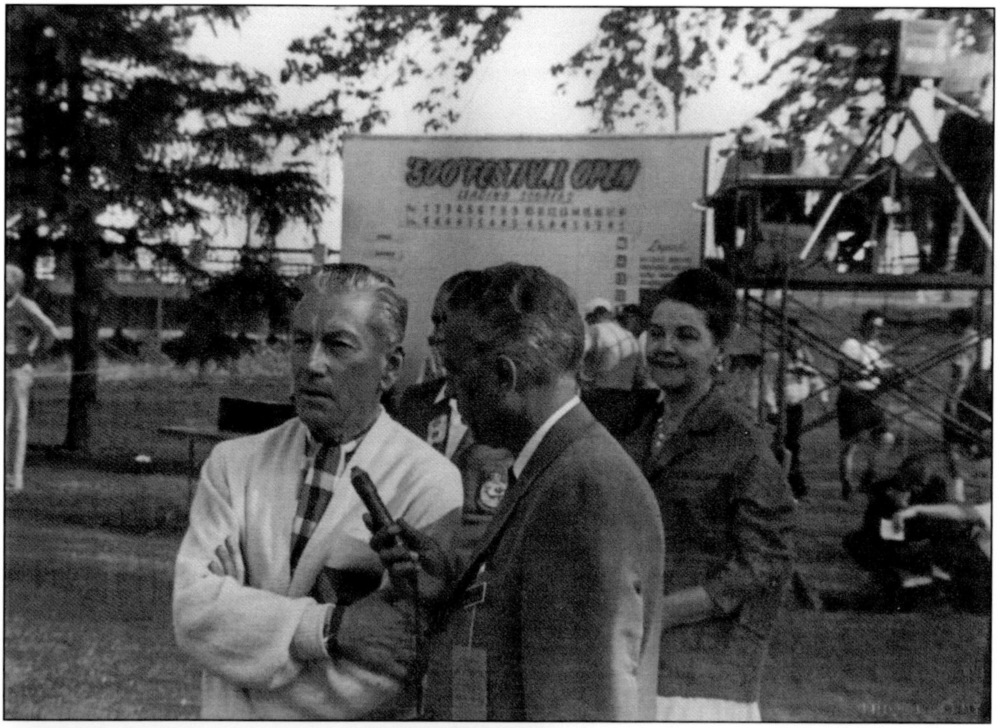

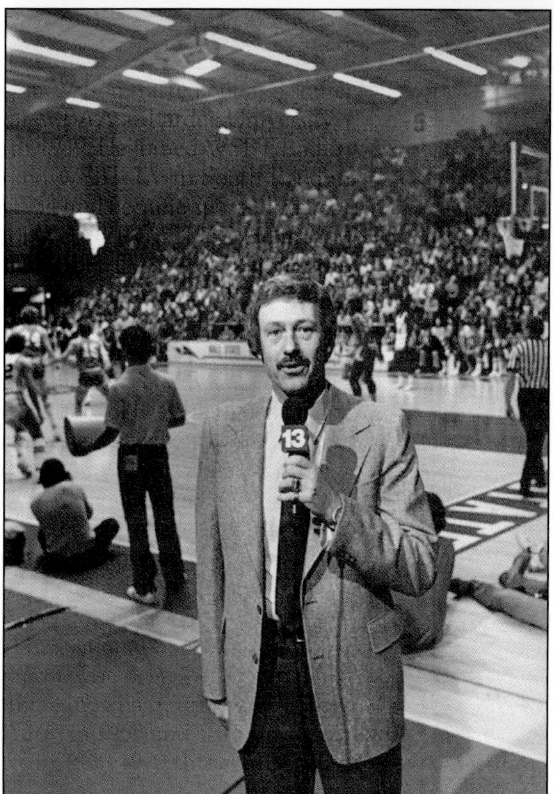

Jim McIntyre was sports director at WISH TV and then moved over to WFBM TV to back up Tom Carnegie. Here, he is interviewing Hoosier composer Hoagy Carmichael at the 500 Festival golf tournament. McIntyre left WFBM TV to become the sports play-by-play man for the Cincinnati Reds. He called the action from 1966 to 1969.

Sports director Don Hein had a busy 33-year run at WTHR TV. Originally from Waverly, Iowa, Hein has covered everyone from Billy Keller to Peyton Manning and everything in between. He was named "Sportscaster of the Year" three times by the Associated Press and United Press International. He is a member of the Indiana Sportscasters/Sportswriters Hall of Fame, the Indiana Associated Press Hall of Fame, and the Indiana Television Pioneers Hall of Fame. Over the years, he accumulated many additional honors for his distinctive style of sports reporting.

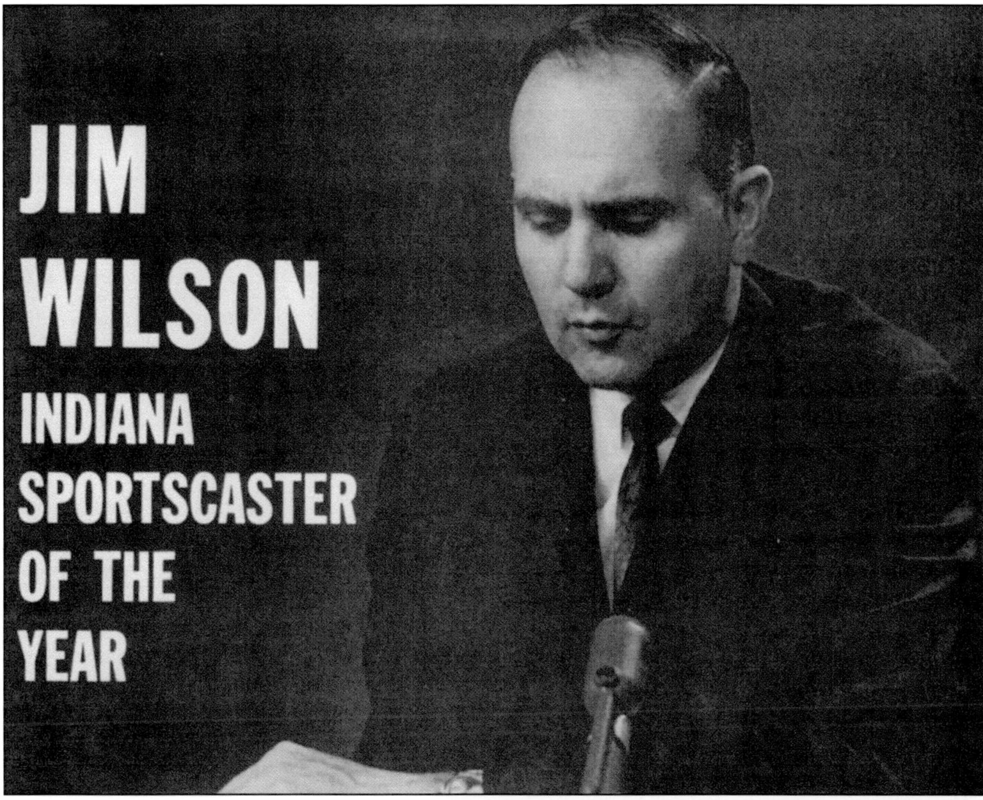

Jim Wilson started out as a staffer at WBAA at Purdue University in 1949. He joined WSBT Radio and WSBT TV in South Bend in 1954 and became sports director of WSBT TV in 1960. He moved to WISH TV as sports director in 1964 and continued in that capacity through the early 1970s. He covered a variety of sports, including the Indianapolis 500 Race qualifications, high school basketball, Indianapolis Indians baseball, and the national clay-court tennis championships. He was an announcer on the MCA TV telecast of the Indianapolis 500 Race for five years and was frequently called upon by CBS to report Indiana sports nationwide. The National Sportscasters and Sportswriters Association named him "Indiana Sportscaster of the Year" in 1968 and again in successive years from 1970 through 1973.

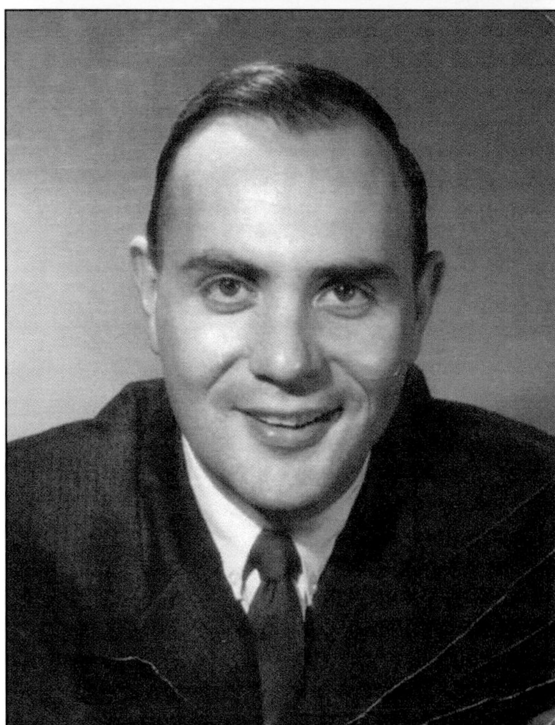

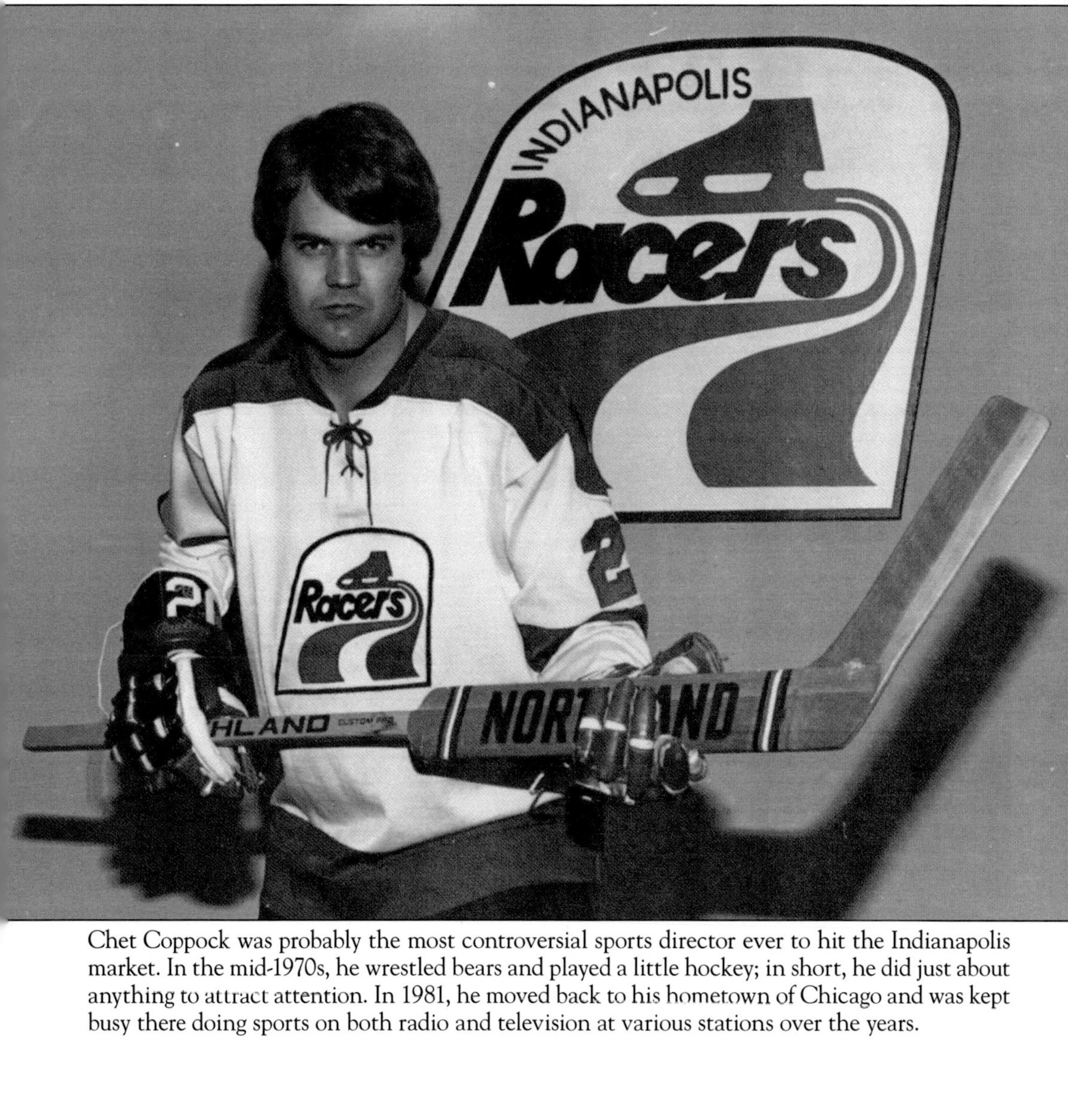

Chet Coppock was probably the most controversial sports director ever to hit the Indianapolis market. In the mid-1970s, he wrestled bears and played a little hockey; in short, he did just about anything to attract attention. In 1981, he moved back to his hometown of Chicago and was kept busy there doing sports on both radio and television at various stations over the years.

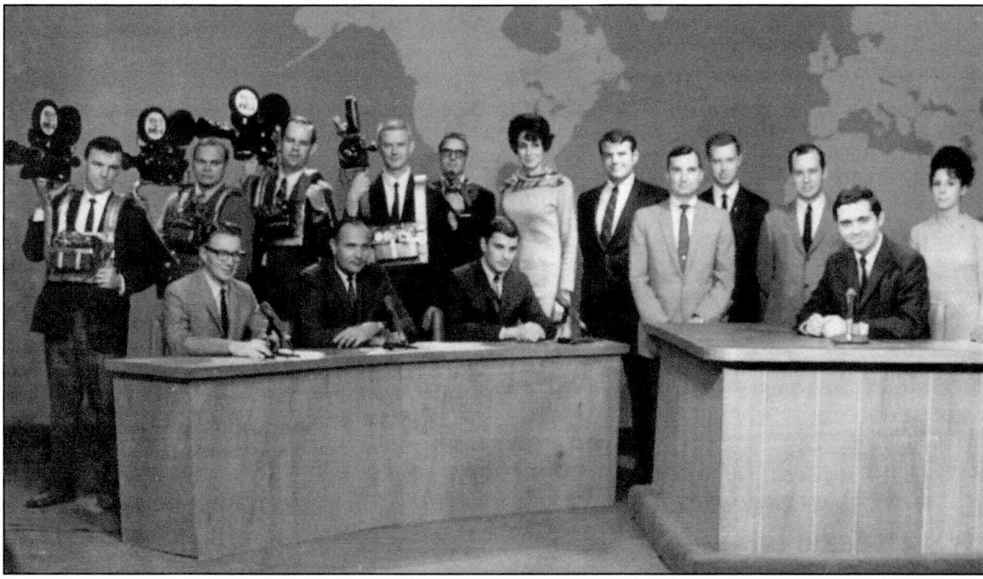

News departments began growing substantially in the 1960s. WISH TV has 15 news staffers in this photograph. The primary team included (seated at desks), from left to right, Stan Wood, weather; Jim Wilson, sports; Rolland Smith, coanchor; and Mike Ahern, coanchor. In the second row are, from left to right, photographers Jerry Brossart, Tim Harlan, Dave Brady, Don McLaughlin, and Don Worsham; reporters Faith Levitt and Bob Gore; news director Lee Giles; reporters Tip Kindle, Ray Gray, and Linda Lupear.

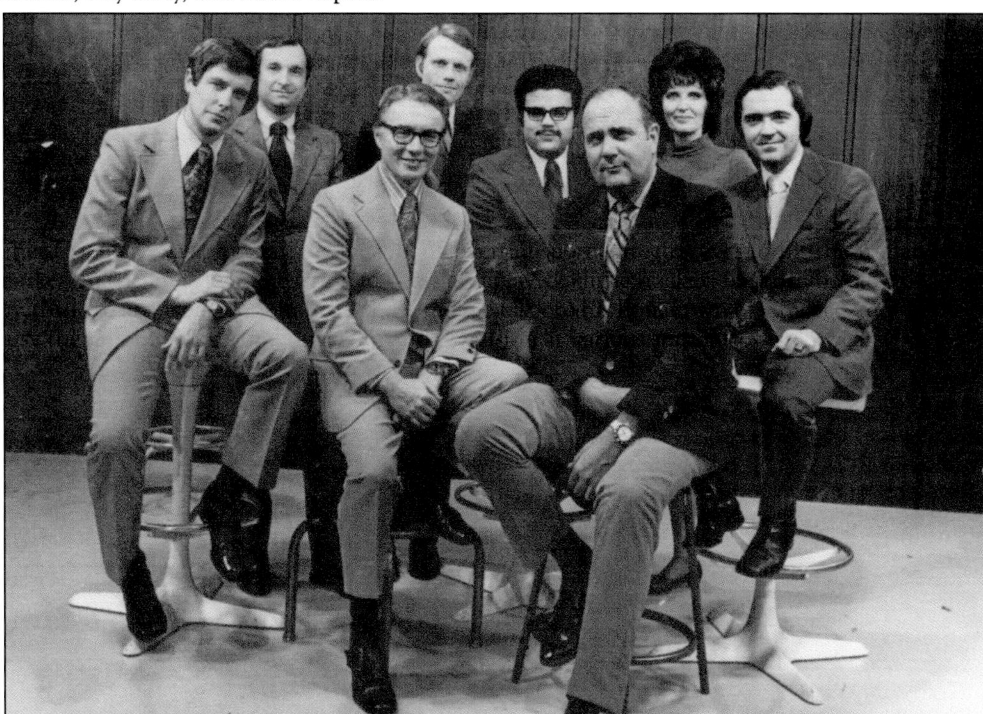

Members of the WISH TV news team in the early 1970s are, from left to right, Jon North, Lee Giles, Stan Wood, Guy Johnson, Dwight Smith, Jim Wilson, Diane Leeds, and Mike Ahern.

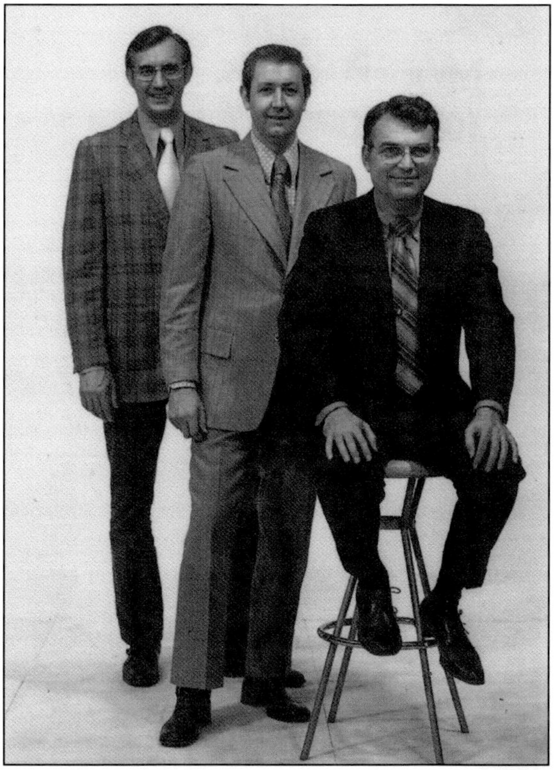

Competing with the WISH TV news team in the 1970s were, from left to right, WFBM TV news anchors Dave Piontek, sports director; Ken Beckley, coanchor; and Howard Caldwell, coanchor.

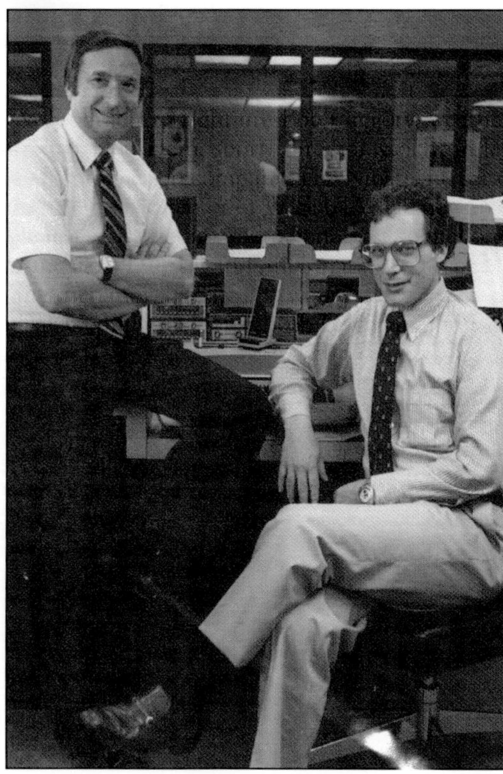

Here is WISH TV news director Lee Giles (left) with reporter Josh Littman. While a student at Butler, Littman was an intern in the WISH TV news department. Lee Giles recognized his ability and hired him, and he became one of the hardest working and brightest reporters in the news department. In 1979, Littman became the anchor for WISH TV's *Nightbeat Live*. He left Indianapolis to go to Atlanta, Georgia, and then to Detroit, Michigan, where he became a top reporter. He unfortunately acquired leukemia and died in Detroit at age 35.

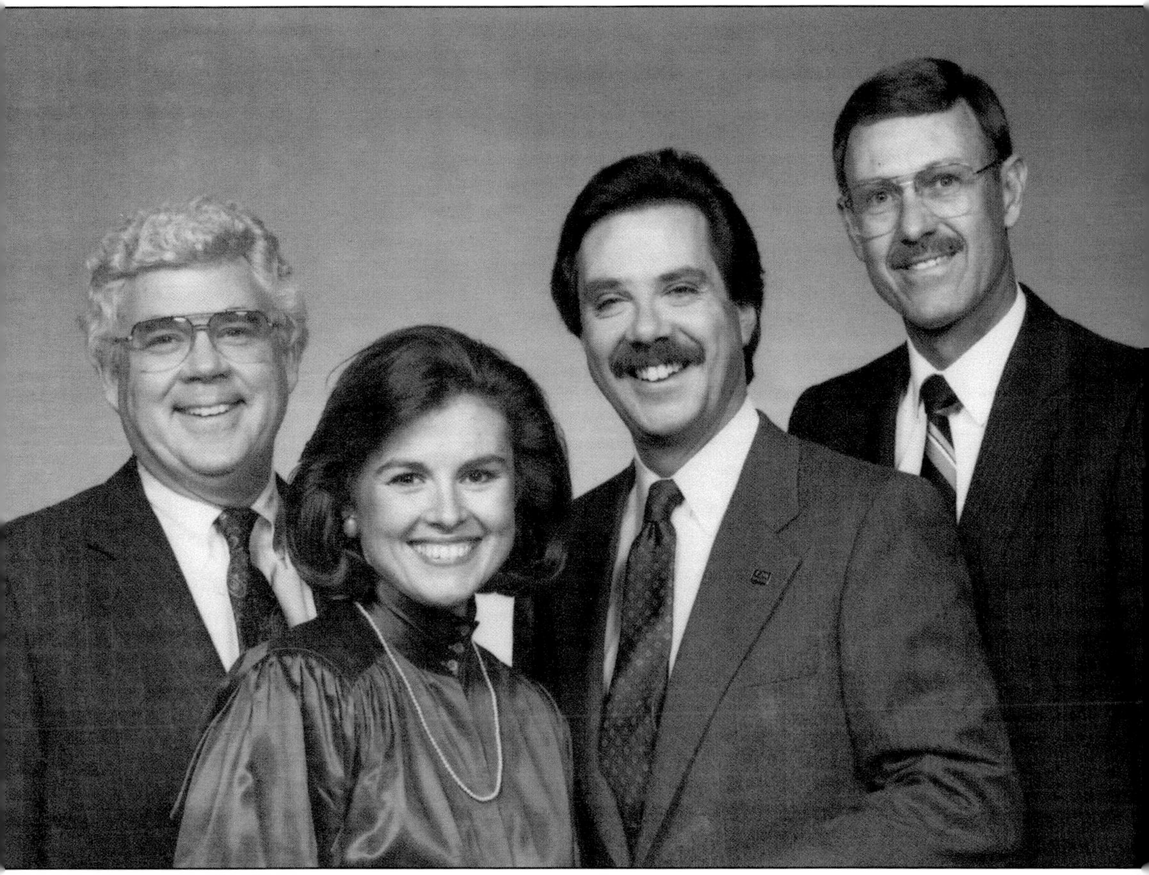

This is an early-1980s news team at WTHR TV. From left to right are Bob Gregory, weather; Betsy Ross, coanchor; Tom Cochrun, coanchor; and Don Hein, sports director. Cochrun left WTHR TV in the early 1990s to start his own media production company, The Nineteenth Star. After several years with this company, he became the news director of WISH TV. He now lives in San Luis Obispo, California, where he is still engaged in media production. Ross coanchored with Cochrun from 1986 to 1990. After anchoring at WLW TV in Cincinnati, she became a sports reporter for WXIX TV in the same city. She then moved to ESPN, where she was a sports anchor. She is president and founder of Game Day Communications. Both Cochrun and Ross are Ball State University graduates.

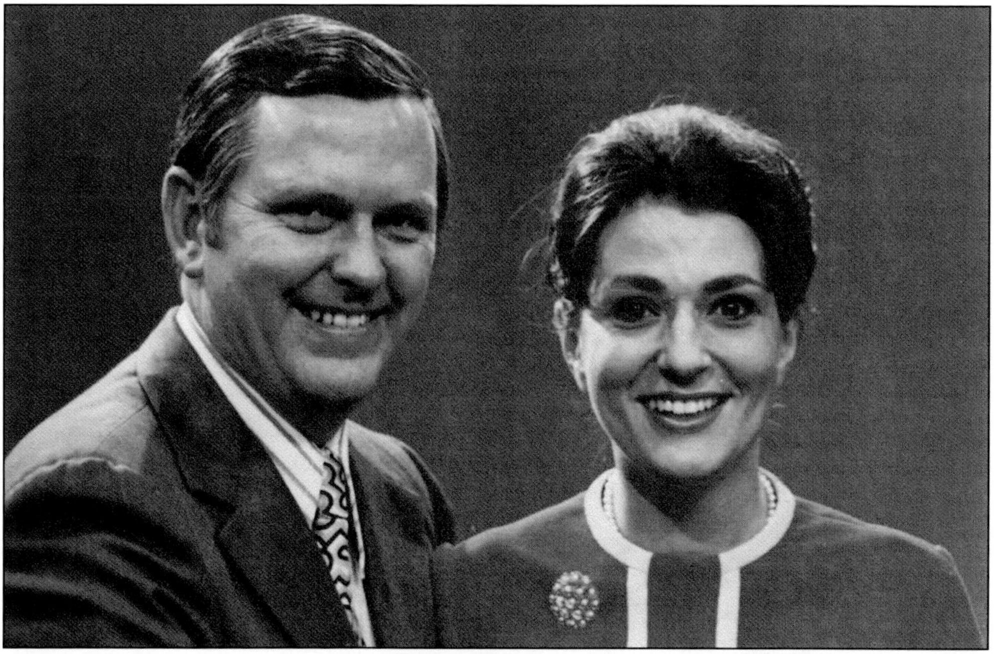

Wally and Natalie Bruner hosted the WISH TV *Morning Show.* The show ran daily from 9:00 a.m. to 10:00 a.m. in the late 1960s. Celebrity guests on the show included such names as Soupy Sales and Henry Morgan. Local politics were also an important part of the show. After its run, which lasted about two years, Wally began producing his *Wally's Workshop* series in Indianapolis; it became a very successful nationally syndicated series.

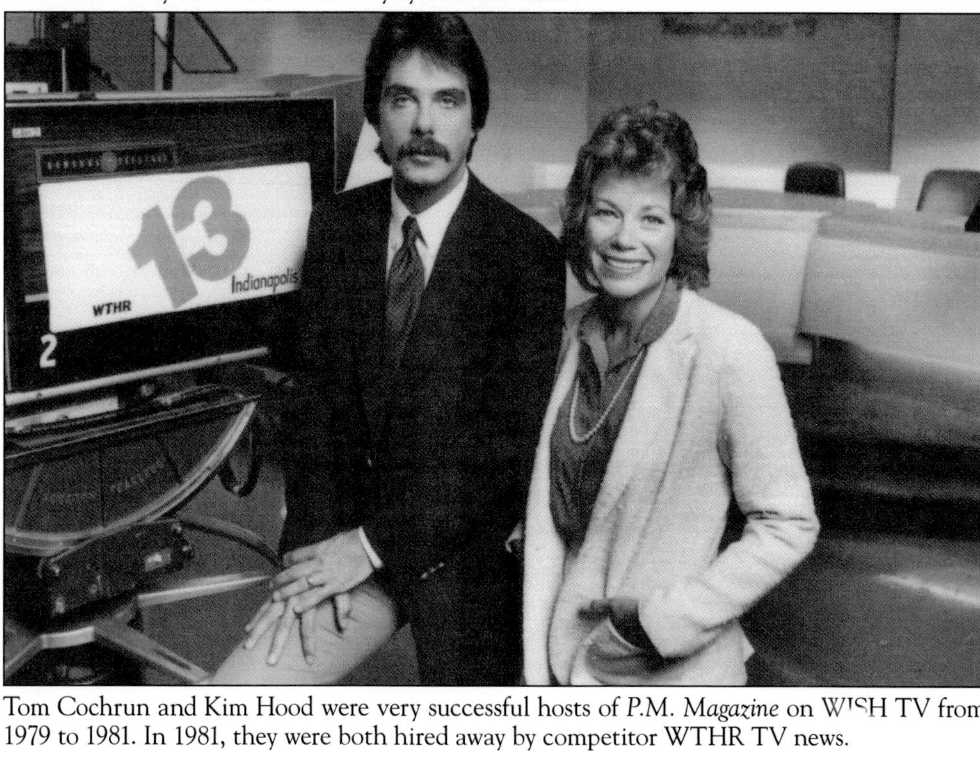

Tom Cochrun and Kim Hood were very successful hosts of *P.M. Magazine* on WISH TV from 1979 to 1981. In 1981, they were both hired away by competitor WTHR TV news.

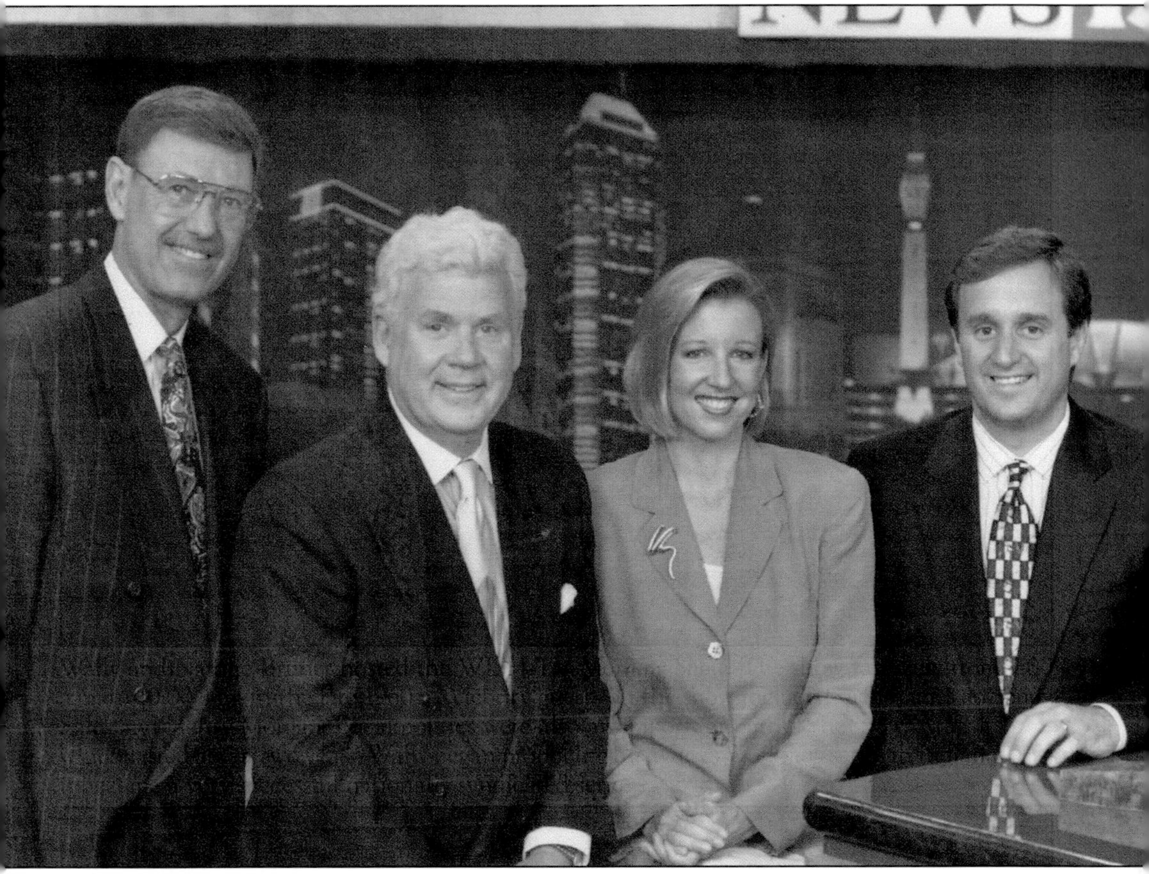

Members of a slightly later news team at WTHR TV are, from left to right, Don Hein, sports; Bob Gregory, weather; Ann Ryder, coanchor; and John Stehr, coanchor. This image is from the late 1990s.

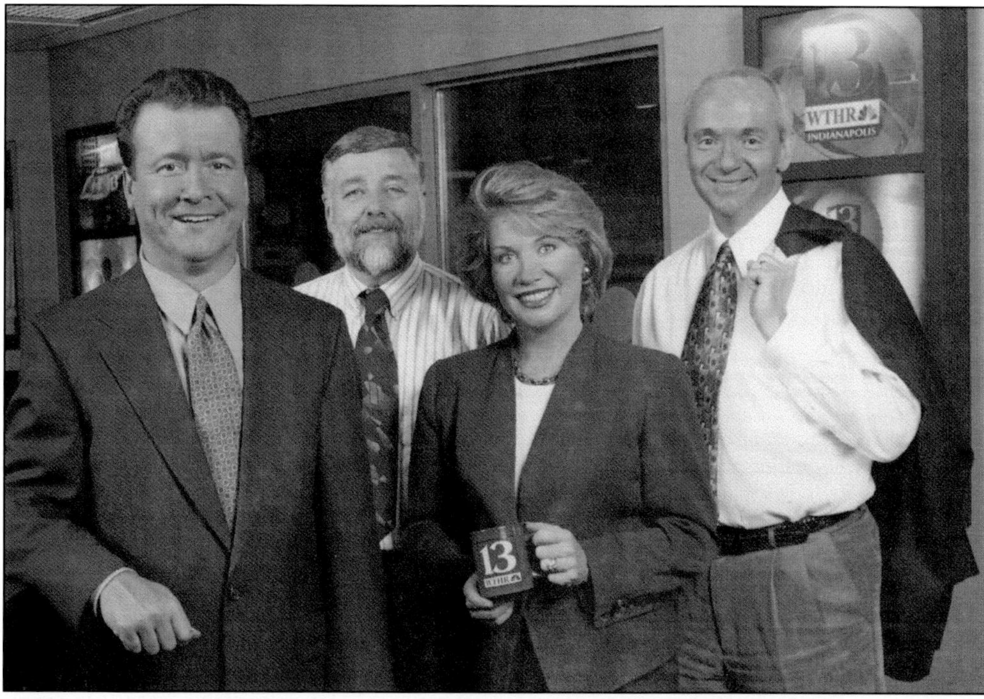

In the early 2000s, WTHR TV began airing *Sunrise* at 6:00 a.m. on a daily basis. The *Sunrise* team included, from left to right, Bruce Kopp, Paul Kasey, Pat Carlini, and Chuck Lofton.

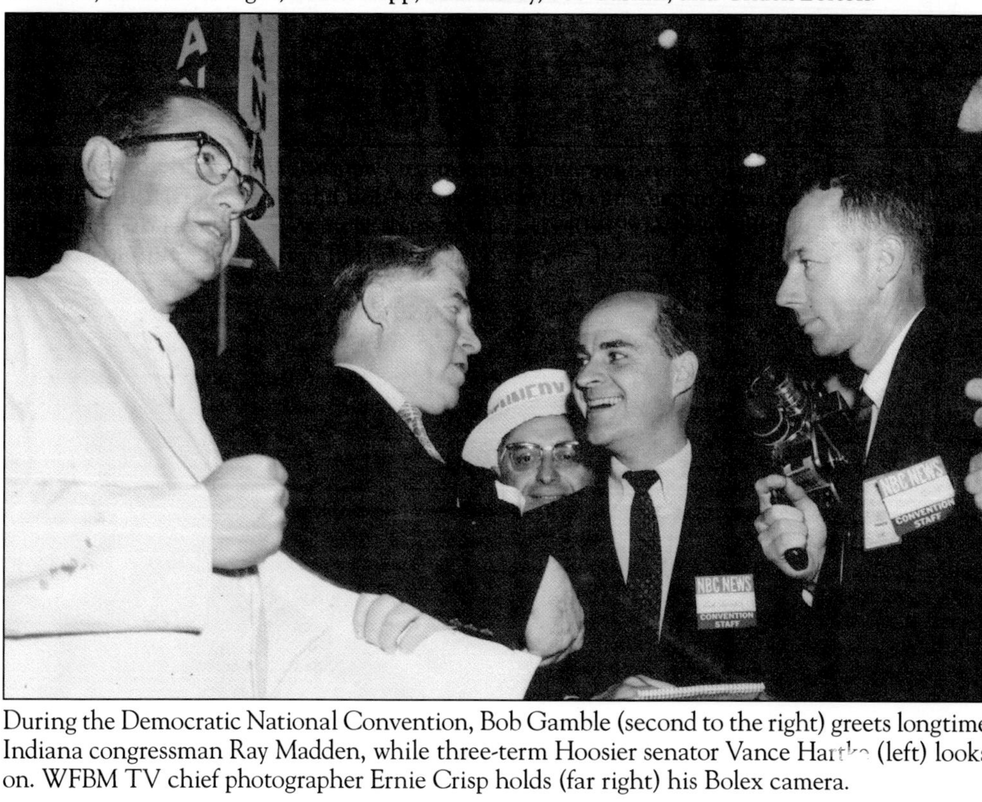

During the Democratic National Convention, Bob Gamble (second to the right) greets longtime Indiana congressman Ray Madden, while three-term Hoosier senator Vance Hartke (left) looks on. WFBM TV chief photographer Ernie Crisp holds (far right) his Bolex camera.

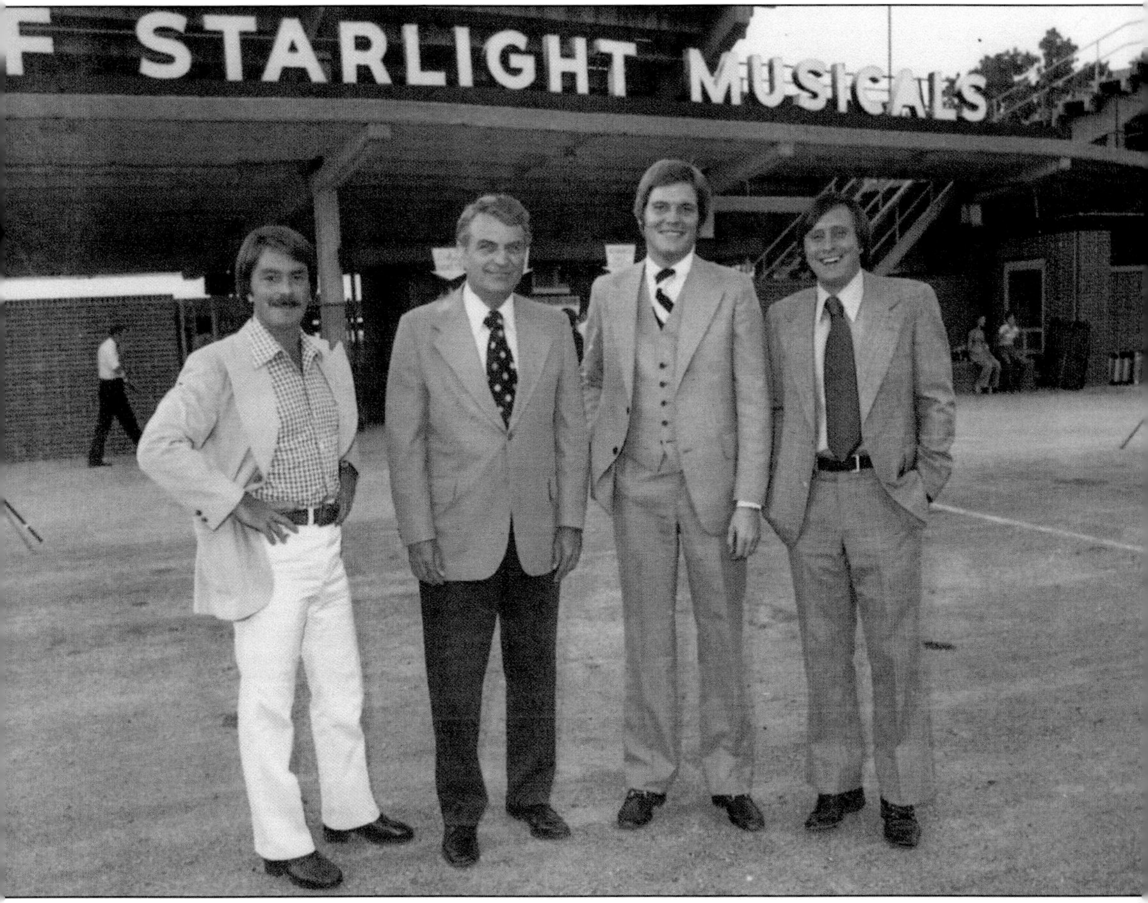

From left to right, WFBM TV news team members Craig Roberts (sports), Howard Caldwell and Clyde Lee (coanchors), and Bob McClain (weather) are seen in front of Starlight Musicals in the mid- to late 1970s. For more than 50 years (1940–1997), Starlight Musicals entertained Indianapolis from its location in the Butler bowl.

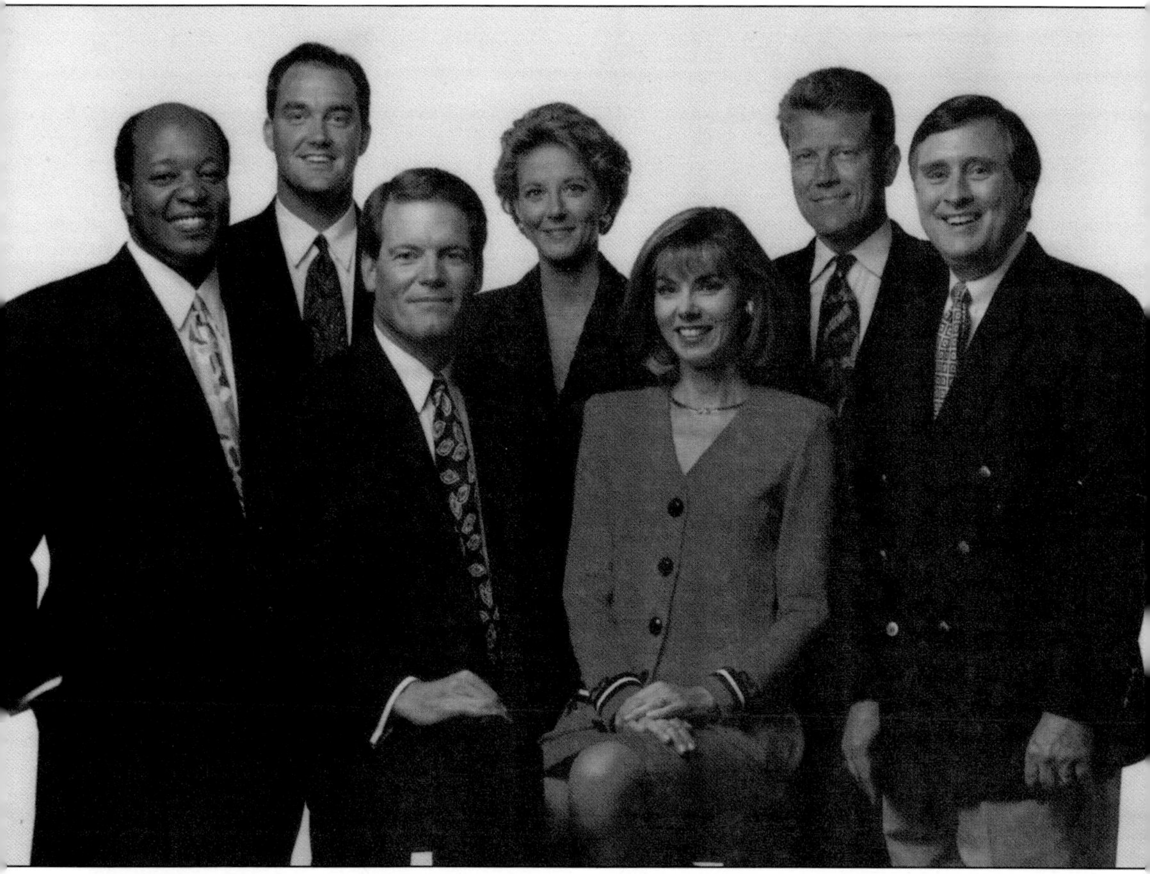

The WFBM TV anchor team is shown in the late 1980s. Clyde Lee replaced Ken Beckley in 1976 as coanchor with Howard Caldwell. Diane Willis joined WRTV in 1987, while Lee and Willis left WRTV in 2001 to start their own communications business in Indianapolis. From left to right are James Adams, Kevin Gregory, Clyde Lee, Barbara Lewis, Diane Willis, Ed Sorenson, and Bob McClain.

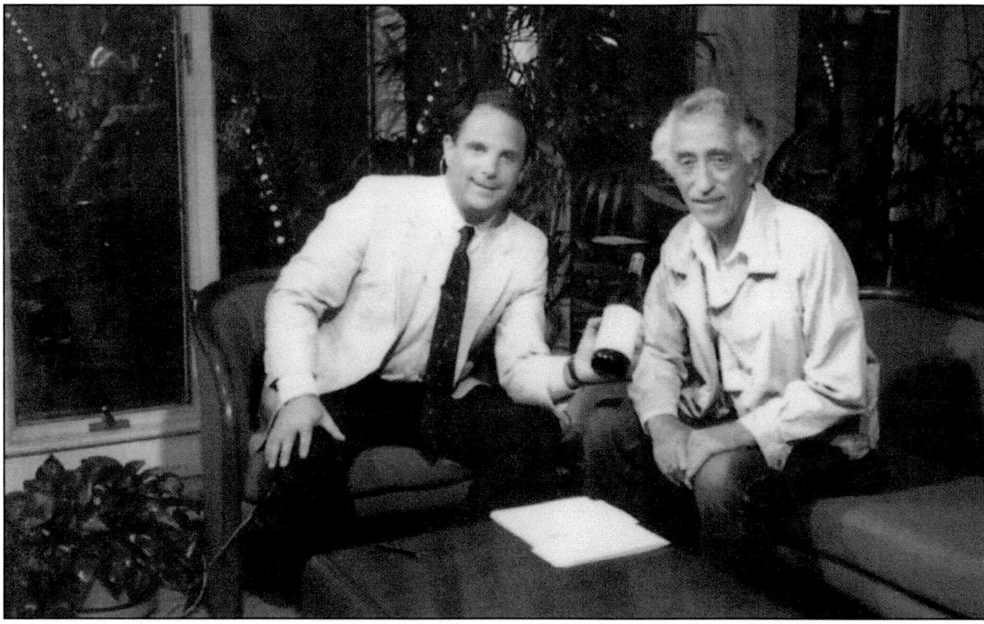

Dick Wolfsie (left) came to Indianapolis in 1982 to do *P.M. Magazine* for WISH TV. His unique blend of wit and humor made him a favorite talk-show host. His dog Barney was featured with him from 1991 to 2004. His book *Mornings with Barney* was a hit, and he wrote 12 other books. He regularly appeared as a member of WISH TV's *Early Show*. He is shown with Pat Paulsen at Union Station, where he hosted a successful talk show on WTHR TV.

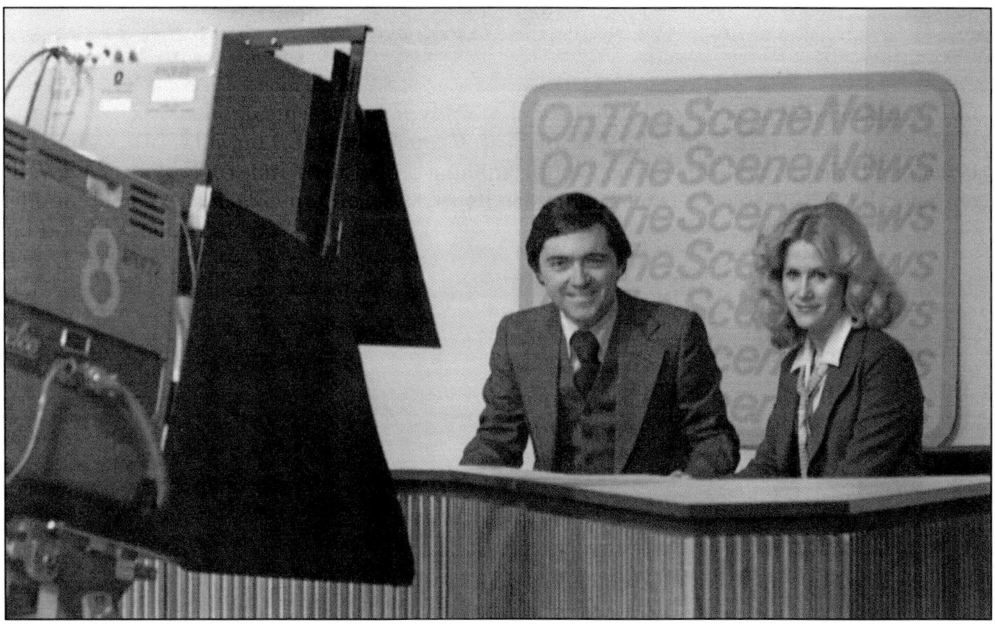

Here are Mike Ahern and Carol Krause as they appeared as coanchors on WISH TV in the mid- to late 1970s. Krause moved to Chicago to take a news anchor position at WBBM TV. She and her husband then moved to Los Angeles, where she did syndicated news reporting and worked for KCBS TV. Her last move was to New York City, where she did some writing and, as a breast cancer survivor, was kept quite busy with speaking engagements.

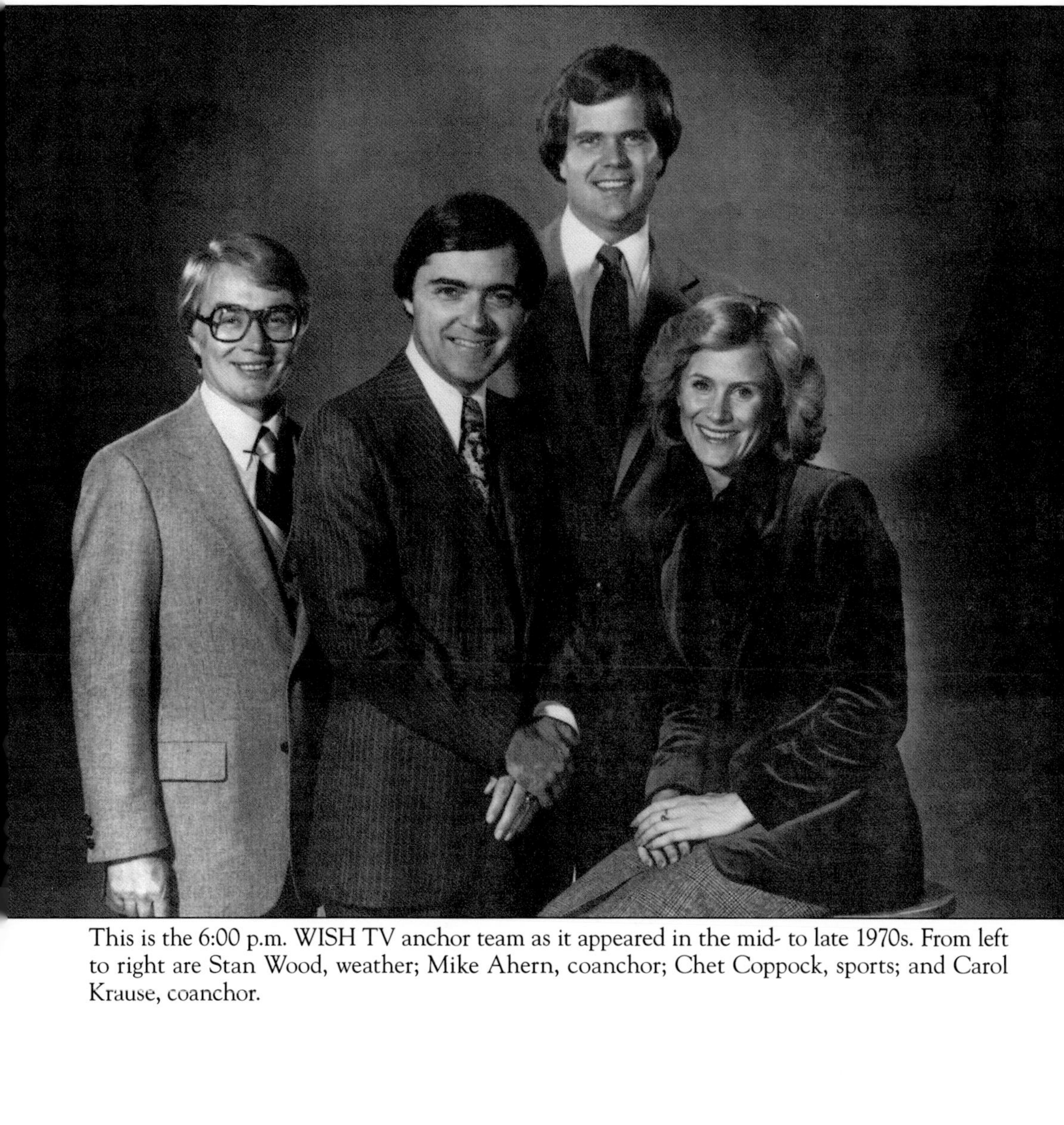

This is the 6:00 p.m. WISH TV anchor team as it appeared in the mid- to late 1970s. From left to right are Stan Wood, weather; Mike Ahern, coanchor; Chet Coppock, sports; and Carol Krause, coanchor.

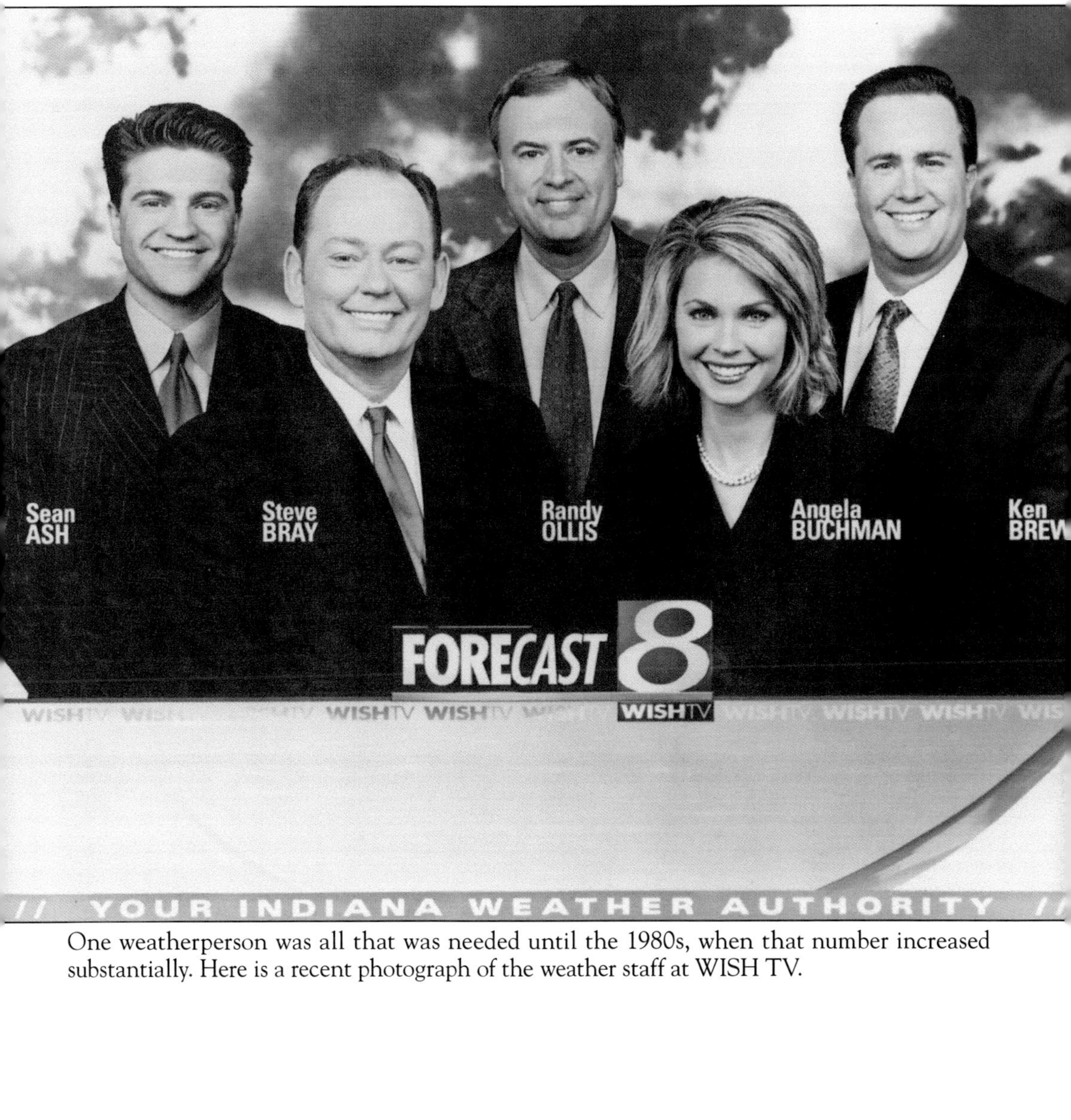

One weatherperson was all that was needed until the 1980s, when that number increased substantially. Here is a recent photograph of the weather staff at WISH TV.

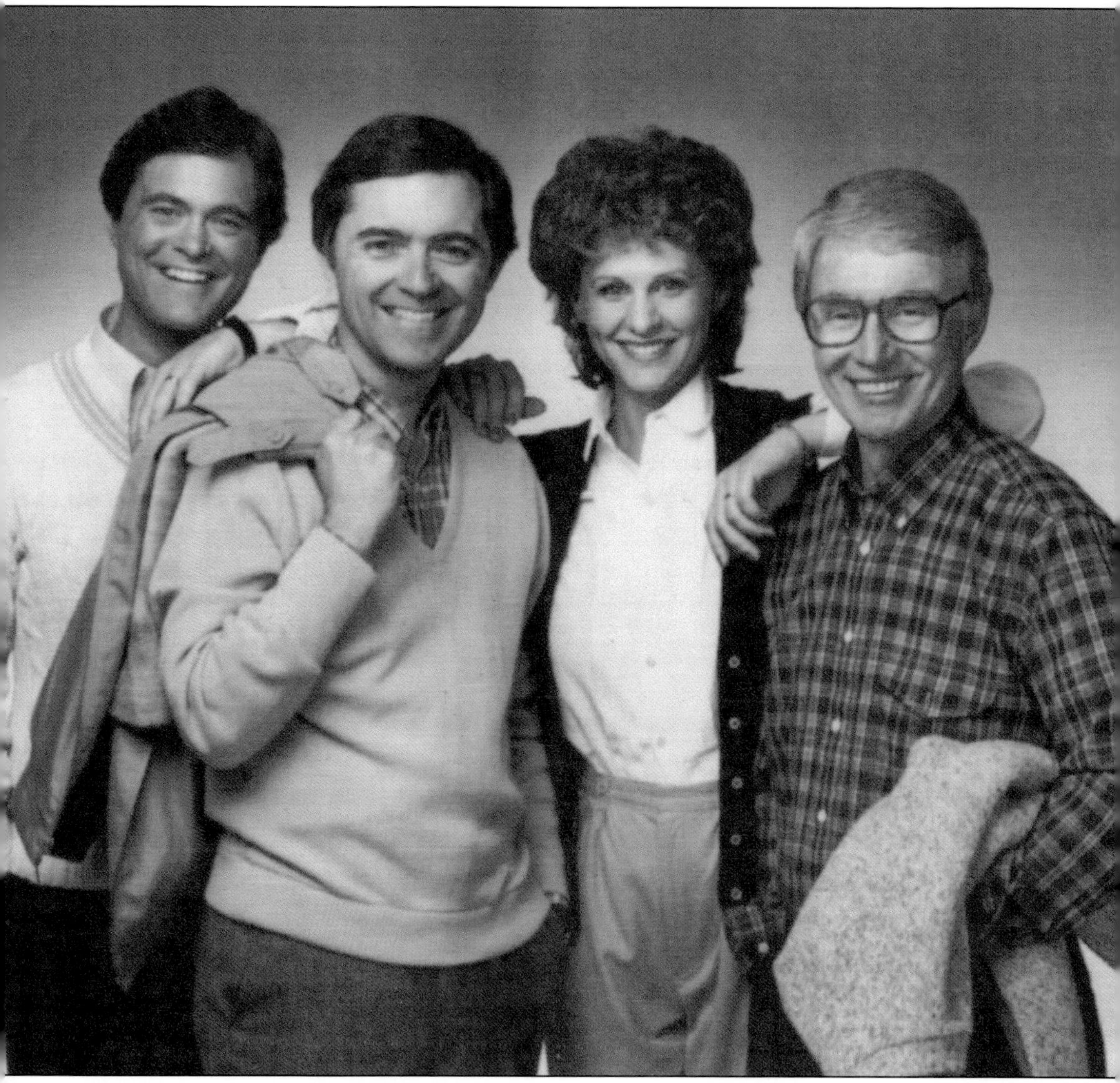

Here are members of the WISH TV news team as they appeared in the early 1980s. Pictured from left to right are Ed Harding, sports; Mike Ahern, coanchor; Debby Knox, coanchor; and Stan Wood, weather.

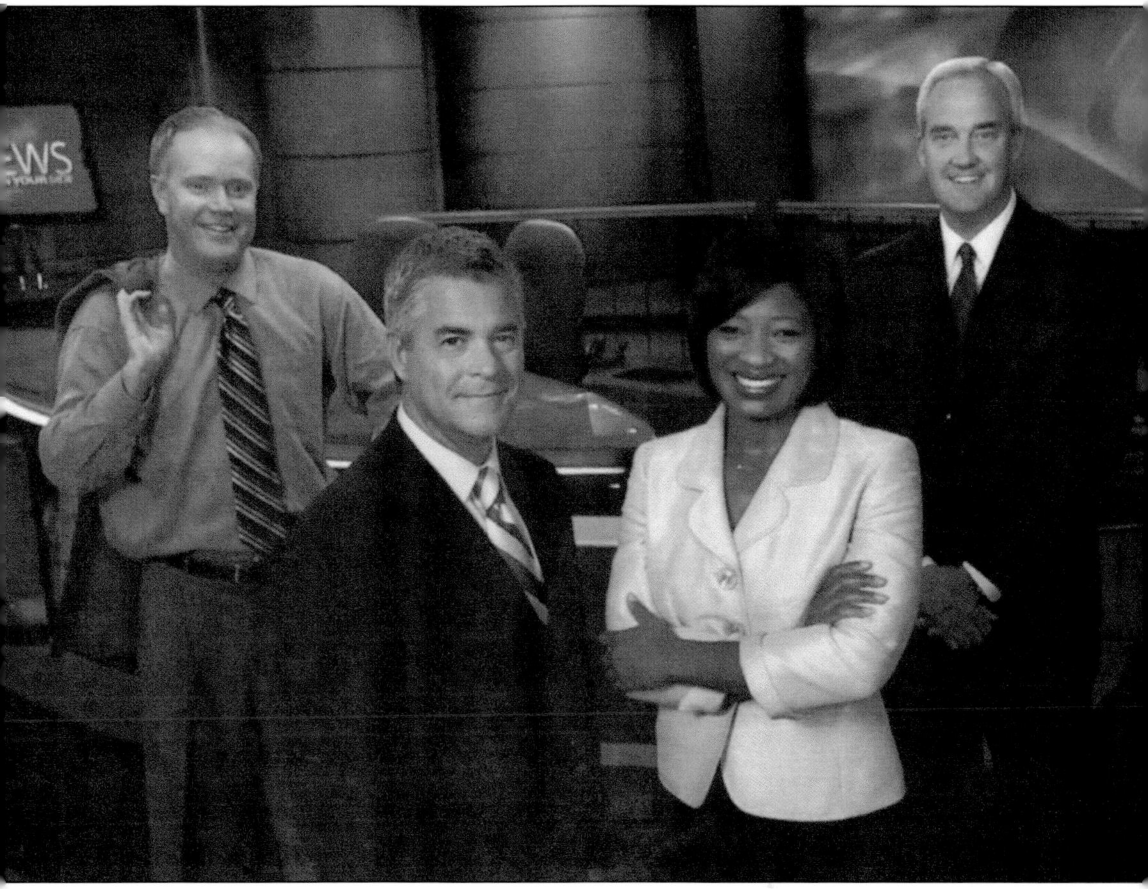

Members of the 2011 WRTV news team are, from left to right, Dave Furst, sports; Todd Connor, coanchor; Ericka Flye, coanchor; and Kevin Gregory, weather.

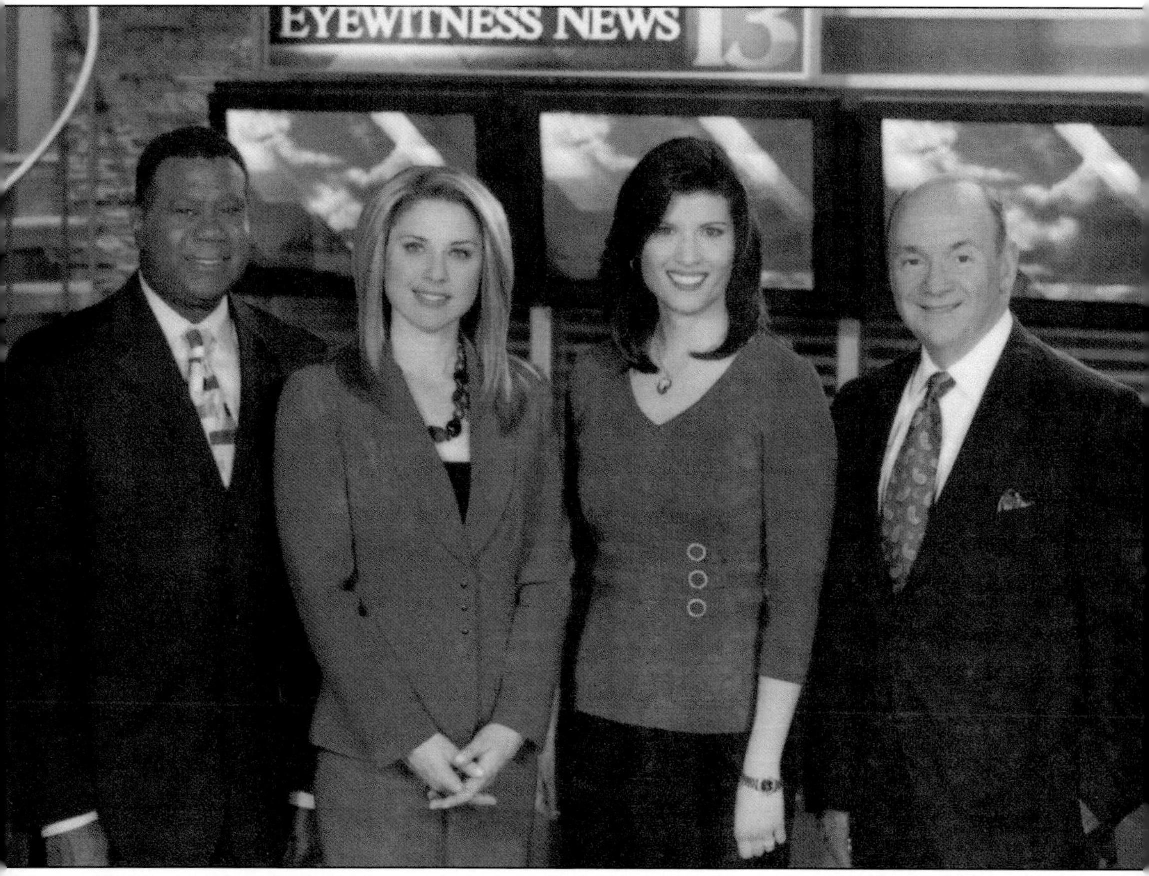

This is the 2011 WTHR TV weather team. Pictured from left to right are Chris Wright, Nicole Misencik, Chikage Windler, and Chuck Lofton.

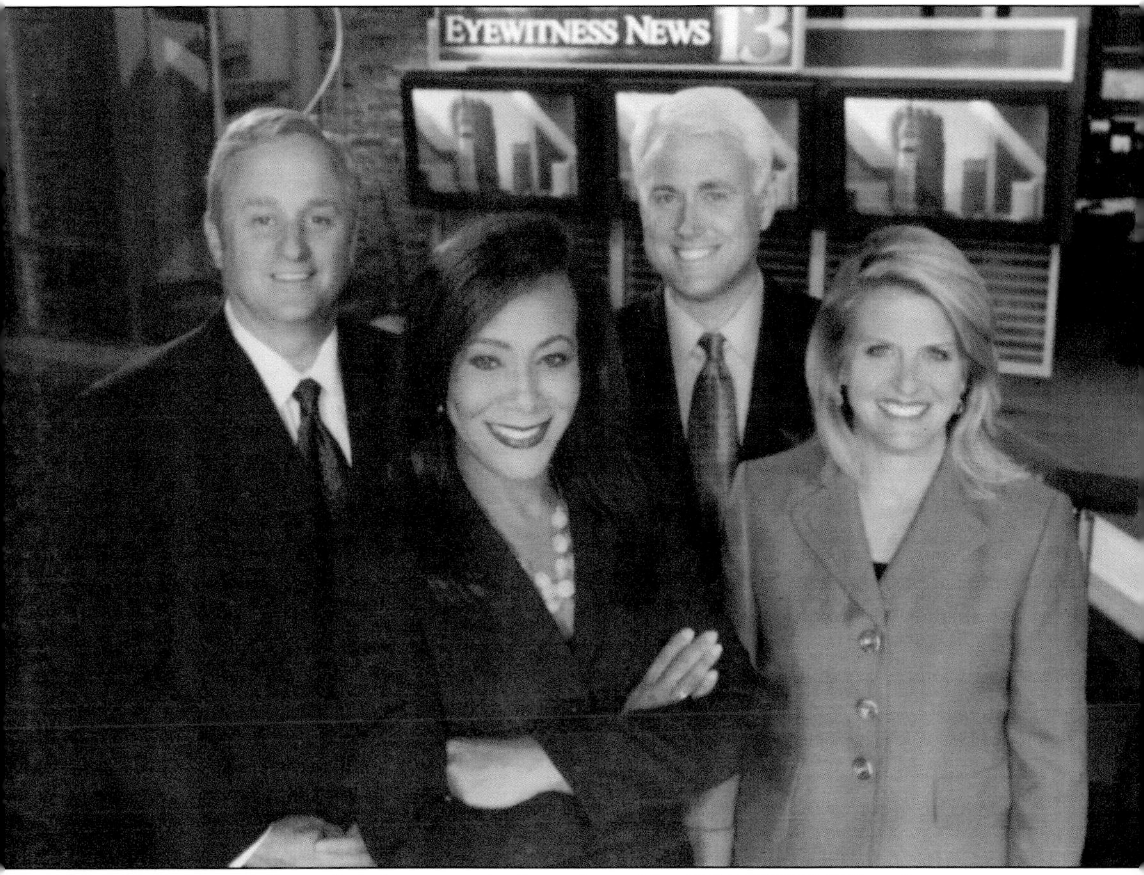

News departments these days feature multiple anchors. The 2011 WTHR TV anchor team poses for a shot here. They are, from left to right, John Stehr, Andrea Moorehead, Scott Swan, and Anna Marie Tiernon.

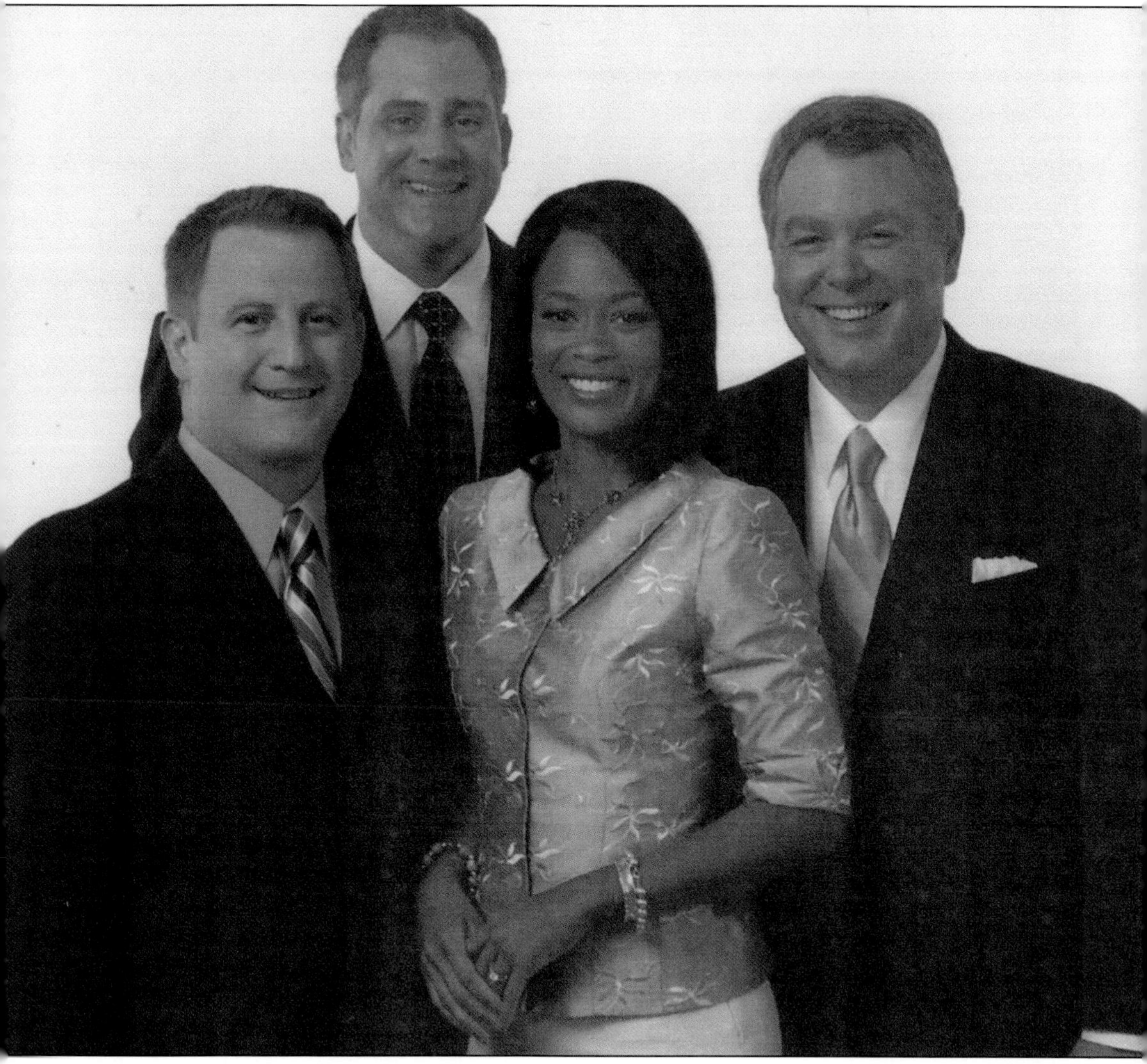

Members of the 2011 WXIN TV anchor team are, from left to right, Chris Hagan, Brian Wilkes, Fanchon Stinger, and Bob Donaldson.

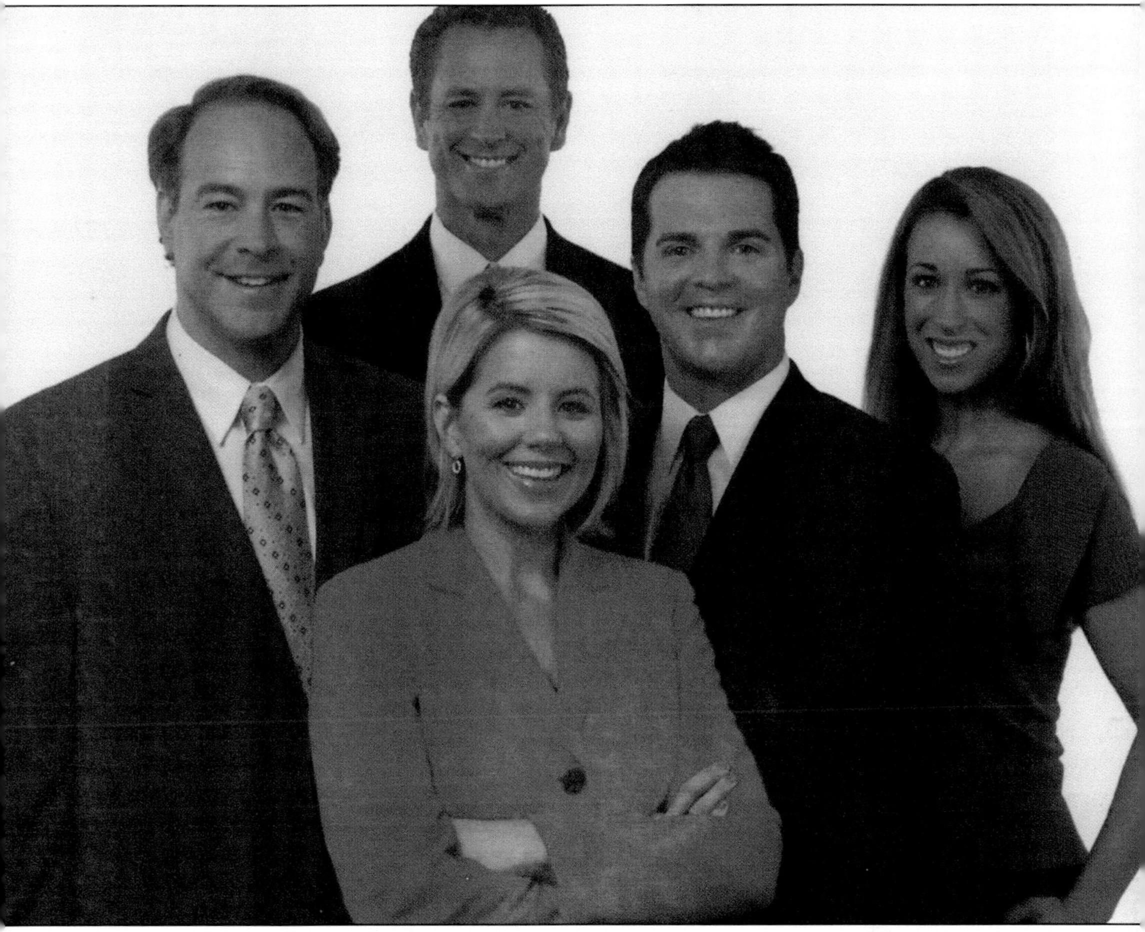

WXIN TV successfully joined other stations in morning news programming. The 2011 morning team consists of, from left to right, Ray Cortopassi, Scott Jones, Angela Ganote, Jim O'Brien, and Jenny Anchondo.

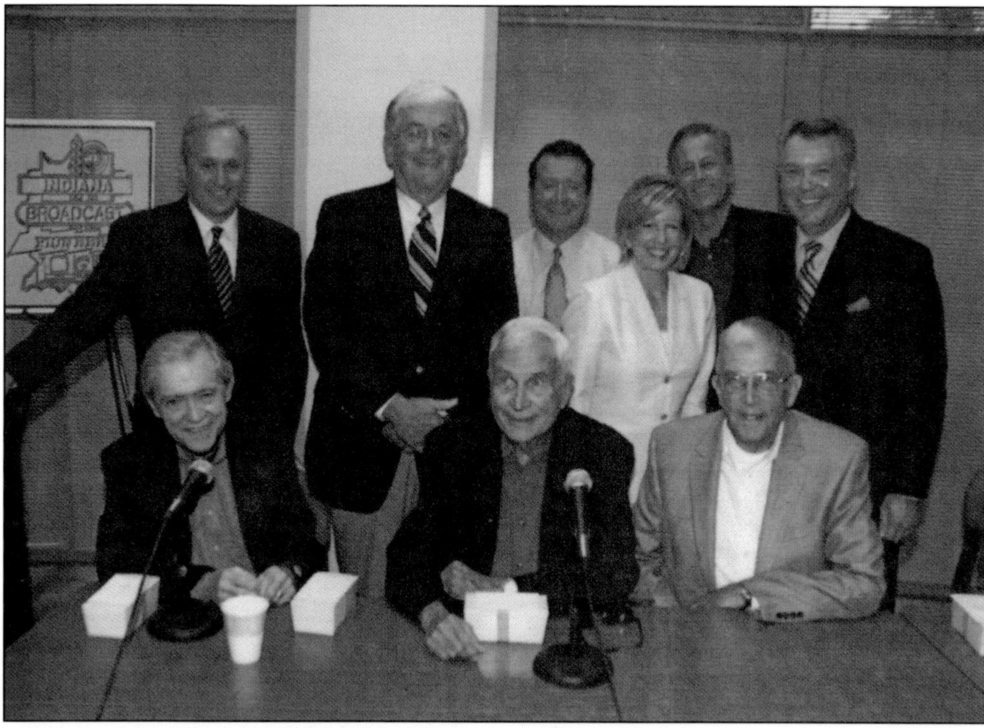

Members of the 2011 Indiana Broadcast Pioneers panel of veteran news anchors are, from left to right, (first row) Mike Ahern, Howard Caldwell, and Dick Florea; (second row) John Stehr, Mike Piggott, Bruce Kopp, Ann Ryder, Ken Owen, and Bob Donaldson.

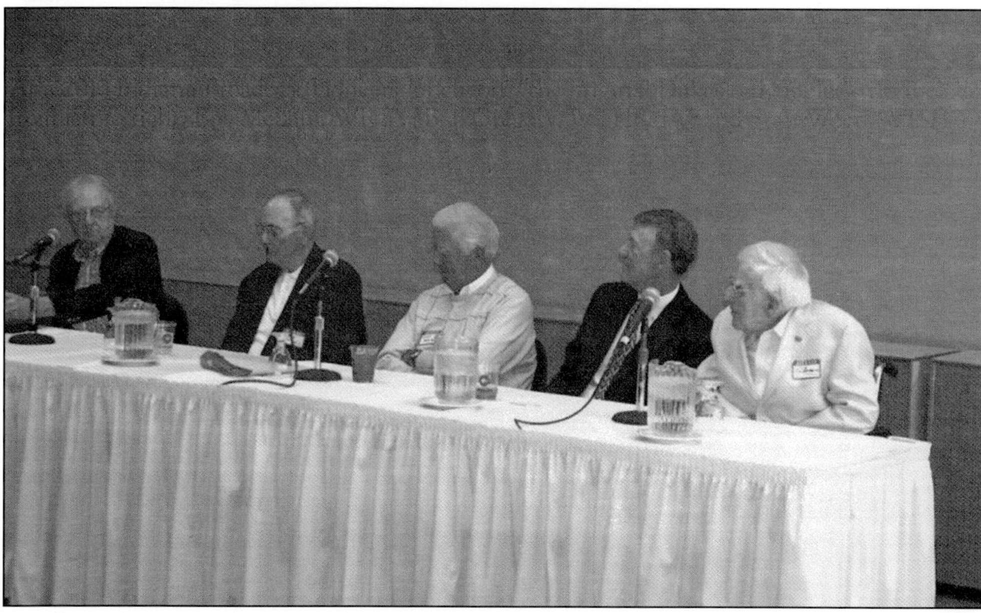

Members of a 2010 Indiana Broadcast Pioneers gathering of veteran sports people are, from left to right, Sam Simmermaker (WCSI Radio), Jim Wilson (sports director of WISH TV), Chuck Marlowe (sports director of WTTV), Don Hein, (sports director of WTHR TV), and Tom Carnegie (WFBM TV/WRTV, the dean of sports directors in Indianapolis).

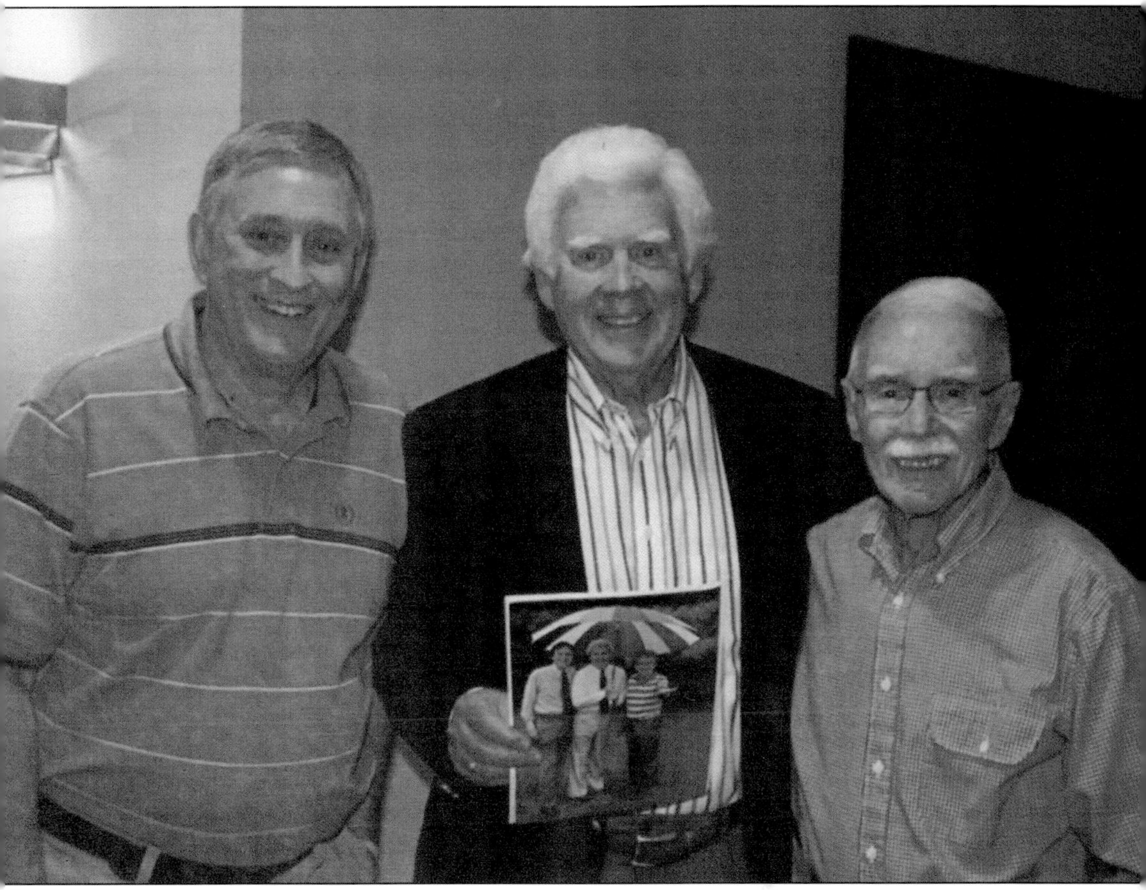

At a 2010 Indiana Broadcast Pioneers luncheon, "the original Dukes of Doppler" trio includes, from left to right, Bob McLain, WRTV; Bob Gregory, WTHR TV; and Stan Wood, WISH TV.

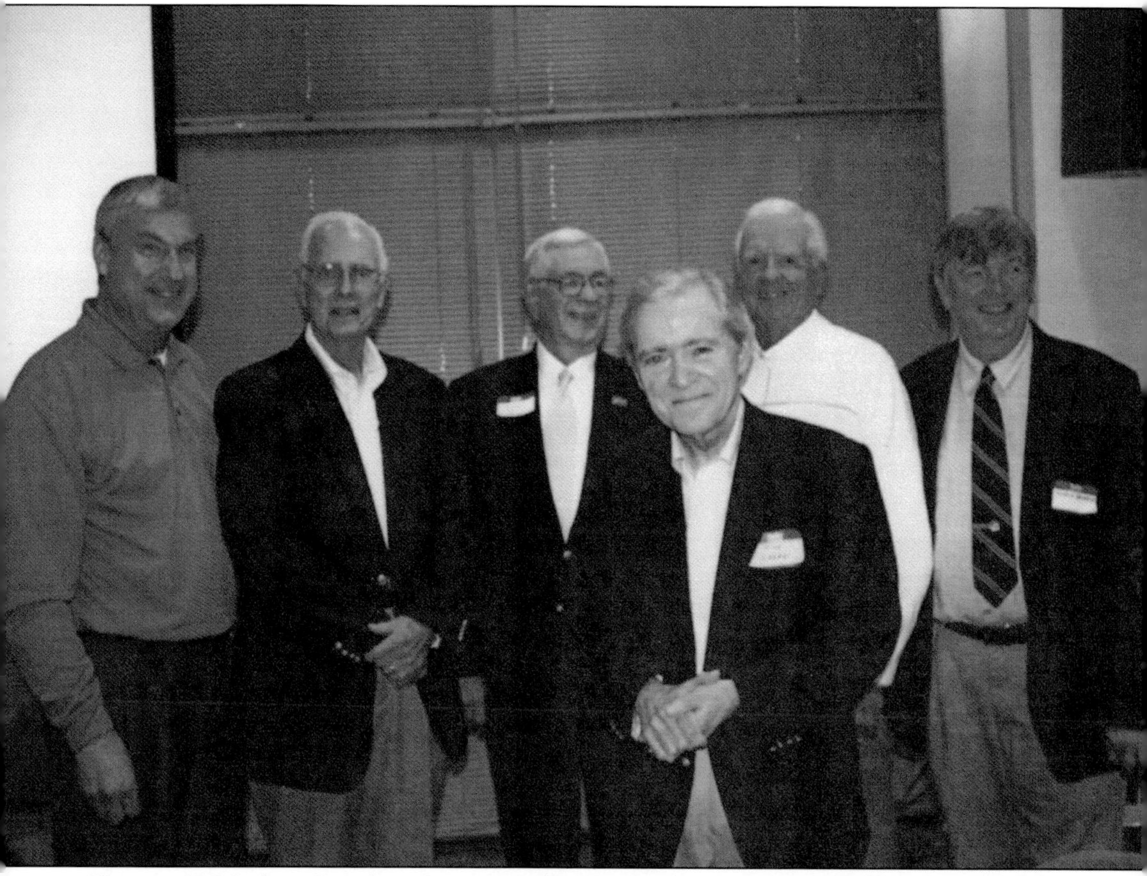

Here is a 2010 Indiana Broadcast Pioneers gathering of Indianapolis 500 Mile Race commentators. From left to right are Bob Jenkins, Doug Zink, Howdy Bell, Mike Ahern, Chuck Marlowe, and Donald Davidson.

INDEX

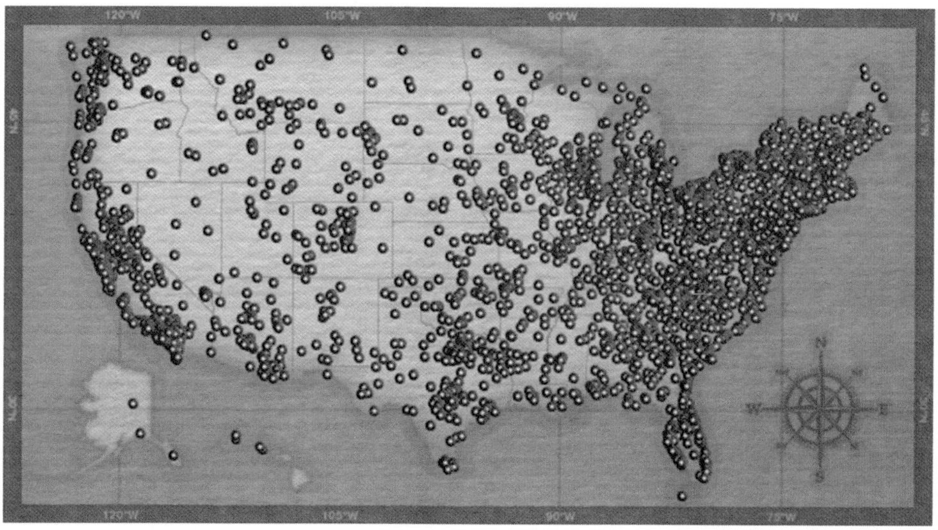